Beiträge zur Altertumskunde

Herausgegeben von
Ernst Heitsch, Ludwig Koenen,
Reinhold Merkelbach, Clemens Zintzen

Band 72

B. G. Teubner Stuttgart und Leipzig

Studies in the Text of Seneca's *Naturales Quaestiones*

Harry M. Hine

B. G. Teubner Stuttgart und Leipzig 1996

Die Deutsche Bibliothek – CIP-Einheitsaufnahme

Hine, Harry M.:
Studies in the text of Seneca's Naturales quaestiones /
Harry M. Hine. – Stuttgart; Leipzig: Teubner, 1996
(Beiträge zur Altertumskunde; 72)
ISBN 3-519-07621-7
NE: GT

Printed in Germany
Druck und Bindung: Röck, Weinsberg

PREFACE

This volume of notes on the text and interpretation of Seneca's *Naturales Quaestiones* is a companion to my Teubner edition of the work. I have not attempted a comprehensive textual commentary, but have discussed passages where I have taken a different view of the text from my predecessors, or where I think more needs to be said to justify the text that others have adopted. The notes were completed in September 1994, and only minor alterations have been possible since then. Although it is not listed in the bibliography, I must acknowledge the constant help I have had from the PHI CD-ROM database of Latin texts. Some of my ideas have been tried out on colleagues at the St Andrews Classics Research Seminar, and I have benefited from the ensuing discussion. Professors R. G. M. Nisbet and W. S. Watt kindly commented in detail on earlier versions of my notes, and I have learnt much from them. But my obstinacy and afterthoughts have intervened, and remaining faults are entirely my own responsibility. Finally, I am grateful to the University of St Andrews for help with the costs of publication.

St Andrews, September 1995

H. M. H.

Contents

BIBLIOGRAPHY

Editions

ed. Ven.[1]	Venetiis, 1490 (editio princeps)
ed. Ven.[2]	Venetiis, 1492
ed. Lips.	Lipsiae, c. 1492/95, per Arnoldum de Colonia
ed. Ven[3]	Venetiis, 1503
Erasmus[1]	Basileae, 1515
Fortunatus	Venetiis, 1522
Erasmus[2]	Basileae, 1529
Curio	Basileae, 1557
Muretus	Romae, 1585
Gruterus	Heidelbergiae, 1593
Lipsius	Antverpiae, 1605
Gronouius	Amstelodami, 1658
Ruhkopf, F. E.	Lipsiae, 1811
Koeler, G. D.	Gottingae, 1819
Bouillet, M. N.	Parisiis, 1830 (Bibliotheca Classica Latina)
Fickert, C. R.	Lipsiae, 1845
Haase, F.	Lipsiae, 1852 (ed. Teubneriana)
Gercke, A.	Lipsiae, 1907 (ed. Teubneriana)
Oltramare, P.	Paris, 1929 (Collection des Universités de France Budé)
Corcoran, T. H.	London, Cambridge, MA, 1971-2 (Loeb)
Codoñer Merino, C.	Madrid, 1979 (Colección Hispánica de Autores Griegos y Latinos)
Vottero, D.	Torino, 1989 (Classici Latini)

Books and Articles

[This bibliography includes only authors referred to in this volume. There is a more complete bibliography of works dealing with the text of the *Naturales Quaestiones* in my edition. Where there is more than one work by the same author, I have used the same system of numbering as is used in the edition.]

Alexander	W. H. Alexander, *Seneca's* Naturales Quaestiones. *The Text Emended and Explained* (*University of California Publications in Classical Philology*, 13.8, 241-332)
André	J. André, review of Codoñer, *Gnomon* 53 (1981), 196-7
Axelson[1]	B. Axelson, *Senecastudien, Kritische Bemerkungen zu Senecas Naturales Quaestiones* (*Lunds Universitets Årsskrift* N. F. Avd. 1, Bd. 29, Nr. 3, Lund, 1933)
Axelson[2]	id., *Neue Senecastudien, Textkritische Beiträge zu Senecas Epistulae morales* (*Lunds Universitets Årsskrift* N. F. Avd. 1, Bd. 36, Nr. 1, Lund, 1939)
Shackleton Bailey	D. R. Shackleton Bailey, 'Notes on Seneca's *Quaestiones Naturales*', *CQ* 29 (1979), 448-56
Bentley	R. Bentley, annotations in his copy of the 1672 Elzivir edition of Seneca (now Cambridge University Library Adv.d.52.8)
Bongarsius	J. Bongarsius, annotations in his copy of Seneca, published by Fickert (see Fickert Vol. II, p. xviii)
Bourgery	A. Bourgery, *Sénèque prosateur* (Paris, 1922)
Brakman[1]	C. Brakman, 'Ad Senecae Naturales Quaestiones', *Hermes* 45 (1910), 37-42
Brakman[2]	id., 'Annaeana', *Mnemosyne* 56 (1928), 139-58
Busa & Zampolli	R. Busa, A. Zampolli, *Concordantiae Senecanae* (Hildesheim, New York, 1975)
Busche	K. Busche, 'Kritische Beiträge zu Senecas Naturales Quaestiones', *RhM* 70 (1915), 568-83

Castiglioni[1]	L. Castiglioni, 'Studi Anneani - IV. Note critiche ai libri delle Questioni Naturali', *RFIC* 49 (1921), 435-55
Castiglioni[2]	id., 'Studi Anneani - IV. Note critiche ai libri delle Questioni Naturali (Continuazione e fine)', *RFIC* 50 (1922), 55-66
Castiglioni[3]	id., 'Studi anneani III. Osservazioni critiche alle "Epistole Morali"', *SIFC* n.s. 2 (1922), 209-62
Castiglioni[4]	id., review of Axelson[1], *Gnomon* 16 (1940), 115-7
Castiglioni[5]	id., 'De naturalium quaestionum codice Veneto Marc. XII 141 (Cod. Lat. 1548)', *Miscellanea Giovanni Galbiati*, 1 (*Fontes Ambrosiani*, 25, Milano, 1951), 167-201
Castiglioni[6]	id., 'Contributo alla storia del testo delle Nat. Quaestiones di Seneca', *AFLC* 18 (1951), 81-102
Charney	B. L. Charney, 'Ellipsis of the verb in Seneca's *Epistulae morales*', *CPh* 38 (1943), 46-8
Codoñer Merino	see under Editions
Cornelissen	J. J. Cornelissen, *Coniectanea Latina* (Deventer, 1870)
Courtney	E. Courtney, 'Conjectures in Seneca's Prose Works', *BICS* 21 (1974), 100-6
DK	H. Diels, W. Kranz, *Die Fragmente der Vorsokratiker*[6] (Zurich, Berlin, 1951-2)
Erasmus[1], Erasmus[2]	see under Editions
Fickert	see under Editions
Friedländer	L. Friedländer, *Darstellungen aus der Sittengeschichte Roms in der Zeit von August bis zum Ausgang der Antonine*, Vol. 2 (Leipzig, 1864)
Fromondus	L. Fromondus, *Scholia ad quaestiones naturales*, appended to the 1632 Antwerp edition of Seneca
Garrod[1]	H. W. Garrod, 'Notes on the *Naturales Quaestiones* of Seneca', *CQ* 8 (1914), 272-81

Garrod[2] id., 'Notes on the *Naturales Quaestiones* of Seneca (Continued)', *CQ* 9 (1915), 39-49

Geist[1] H. Geist, *De L. Annaei Senecae Naturalium quaestionum codicibus* (Bamberg, 1914)

Geist[2] id., 'Senecas Naturales quaestiones und Roger Bacos Opus maius', *Blätter für das Bayerische Gymnasialschulwesen* 52 (1916), 178-84

Gemoll[1] W. Gemoll, review of Schultess[2], *WKPh* 5 (1888), 1481-3

Gemoll[2] id., *Kritische Bemerkungen zu lateinischen Schriftstellern* (Progr. Liegnitz, 1890)

Gercke see under Editions

Gercke[1] A. Gercke, *Seneca-Studien*, *JKPh* Suppl. 22.1 (Leipzig, 1895; Hildesheim, 1971)

Gercke[2] id., *Studia Annaeana* (Greifswald, 1900)

Gercke-Norden A. Gercke, E. Norden, *Einleitung in die Altertumswissenschaft*, I (Leipzig, Berlin, 1910)

Gertz[1] M. C. Gertz, *Studia critica in L. Annaei Senecae Dialogos* (Copenhagen, 1874)

Gertz[2] id., 'Emendationes Annaeanae', *Mélanges Graux* (Paris, 1884), 353-79

Goodyear[1] F. R. D. Goodyear, *Incerti auctoris Aetna* (Cambridge, 1965)

Goodyear[2] id., review of Hine[6], *CR* 34 (1984), 44-5

Goodyear[3] id., *The* Annals *of Tacitus, Books 1-6*, I (Cambridge, 1972)

Goodyear[4] id., *Papers on Latin Literature*, ed. K. M. Coleman et al. (London, 1992)

Gronouius see under Editions

Gronouius[1] *Joh. Fred. Gronovii ad L. et M. Annaeos Senecas notae* (Leiden, 1649)

Gronouius[2] *Joh. Fred. Gronovii ad L. et M. Annaeos Senecas notae* (Amsterdam, 1658)

Gronouius[3] id., annotations in his copy of the 1619 Leiden edition of Seneca (now Leiden, Bibliotheek der Rijksuniversiteit, 763.D3)

Gronouius[4] *Jo. Fr. Gronovii Notae in L. Ann. Senecae Naturales Quaestiones*, primus ed. C. R. Fickert, I-II (Breslau, 1846, 1848)

Gross	N. Gross, *Senecas Naturales Quaestiones: Komposition, naturphilosophische Aussagen und ihre Quellen* (*Palingenesia*, 27, Stuttgart, 1989)
Gruterus	see under Editions
Gruterus[1]	J. Gruterus, *Suspicionum libri IX* (Wittenberg, 1591)
Gruterus[2]	id., *Animadversiones in L. Annaei Senecae Opera* (Heidelberg, 1594)
Haase	see under Editions
Haupt[1], Haupt[2]	M. Haupt, *Opuscula*, Vol. 2, Vol. 3 (Leipzig, 1876; Hildesheim, 1967)
Havet	L. Havet, *Manuel de critique verbale appliquée aux textes latins* (Paris, 1911)
Hine[1]	H. M. Hine, 'Livy's Judgement on Marius (Seneca, *Natural Questions* 5.18.4; Livy, *Periocha* 80)', *LCM* 3 (1978), 83-7
Hine[2]	id., 'Escorial MS. O III 2 and Related Manuscripts of Seneca's *Natural Questions*', *CQ* 28 (1978), 296-311
Hine[3]	id., 'The Manuscript Tradition of Seneca's Natural Questions: Some Manuscripts Related to Z', *Prometheus* 5 (1979), 63-72
Hine[4]	id., 'The Manuscript Tradition of Seneca's *Natural Questions*', *CQ* 30 (1980), 183-217
Hine[5]	id., 'Seneca and Anaxagoras on Snow', *Hermes* 108 (1980), 503
Hine[6]	id., *An Edition with Commentary of Seneca* Natural Questions, *Book Two* (New York, 1981)
Hine[7]	id., review of Codoñer, *CR* 32 (1982), 30-2
Hine[8]	id., 'Obiter dicta on dicto = dictito', *LCM* 7 (1982), 29
Hine[9]	id., '*Natural Questions*', ap. L. D. Reynolds, ed., *Texts and Transmission: a Survey of the Latin Classics* (Oxford, 1983), 376-8
Hine[10]	id., 'The Date of the Campanian Earthquake: A.D. 62 or A.D. 63, or both?', *AC* 53 (1984), 266-9

Hine[11] id., 'Seneca and Anaxagoras in Pseudo-Bede's De mundi celestis terrestrisque constitutione', *Viator* 19 (1988), 111-27

Hine[12] 'The Manuscript Tradition of Seneca's *Natural Questions*: Addenda', *CQ* 42 (1992), 558-62

Hine[13] 'Notes on Seneca's *Natural Questions*', ap. H. D. Jocelyn, H. Hurt, edd., *Tria Lustra: Essays and Notes Presented to John Pinsent* (*Liverpool Classical Papers*, 3, Liverpool, 1993), 311-4

Holl K. Holl, *Die Naturales quaestiones des Philosophen Seneca* (Diss. Berlin, 1935)

Housman[1] A. E. Housman, 'Tunica retiarii', *CR* 18 (1904), 395-8

Housman[2] id., '*Praefanda*', *Hermes* 66 (1931), 402-12

Housman[3] id., 'Fragmenta poetarum', *CR* 49 (1935), 166-8

Housman[4] *The Classical Papers of A. E. Housman*, ed. J. Diggle, F. R. D. Goodyear (Cambridge, 1972)

H-Sz J. B. Hofmann, A. Szantyr, *Lateinische Syntax und Stilistik*[2] (München, 1972)

Jan L. von Jan, 'Neueste Literatur des Seneca', *NJPhP* 31 (1841), 261-72

Knecht D. Knecht, review of Codoñer Merino, *AC* 52 (1983), 386-8

Kroll[1] W. Kroll, ap. Gercke

Kroll[2] id., 'Randbemerkungen', *RhM* 66 (1911), 161-75

K-S R. Kühner, C. Stegmann, *Ausführliche Grammatik der lateinischen Sprache*, II, *Satzlehre*[5] (Darmstadt, 1976)

Larisch[1] B. Larisch, *De Senecae naturalium quaestionum codice Leidensi Voss. et locis illorum librorum a Vincentio Bellovacensi excerptis*, I (Diss. Breslau, 1865)

Larisch[2] id., *Ein Beitrag zur Kritik des ersten Buches von den 'Naturales Quaestiones' des Seneca* (Progr. Sagan, 1870)

Larisch[3] id., *Ein Beitrag zur Kritik des ersten Buches von den Naturales Quaestiones des Seneca* (Progr. Patschkau, 1874)

Larisch[4] id., *Ein Beitrag zur Kritik des zweiten Buches von Senecas Naturales Quaestiones* (Progr. Patschkau, 1879)

Leo F. Leo, ap. Gercke

Lipsius see under Editions

Lipsius[1] J. Lipsius, *Saturnalium sermonum libri duo* (Antwerp, 1582)

Löfstedt[1] E. Löfstedt, Syntactica: *Studien und Beiträge zur historischen Syntax des Lateins*, 1^2 (Lund, 1956), 2 (Lund, 1933)

Löfstedt[2] id., *Coniectanea: Untersuchungen auf dem Gebiete der antiken und mittelalterlichen Latinität*, 1 (Uppsala, Stockholm, 1950)

L & S C. T. Lewis, C. Short, *A Latin Dictionary* (Oxford, 1879)

Madvig[1], Madvig[2], Madvig[3]
J. N. Madvig, *Adversaria Critica*, 1-3 (Copenhagen, 1871-4)

Michaelis[1] H. C. Michaelis, 'Notae ad L. A. Senecae Naturalium Quaestionum libros VII, editos a Friderico Haasio, et collatos cum codice no. 69 Vossiano ex Bibliotheca Lugduno-Batava', *Philologus* 8 (1853), 445-60

Michaelis[2] id., 'Notae ad Senecae Naturalium Quaestionum Lib. III-VII', *Philologus* 9 (1854), 324-45

Mueck H. Mueck, *Observationes criticae grammaticae in L. Annaei Senecae scripta philosophica* (Diss. Marburg, 1890)

C. F. W. Müller C. F. W. Müller, 'Zu beiden Seneca und anderen lateinischen Prosaikern', *NJPhP* 93 (1866), 483-503

G. Müller G. Müller, *De Senecae quaestionibus naturalibus* (Diss. Bonn, 1886)

J. Müller[1] J. Müller, 'Kritische Studien zu Seneca De beneficiis und De clementia', *SAWW* 127.3 (1892)

J. Müller[2] id., 'Über die Originalität der Naturales Quaestiones Senecae', *Fest-Gruss aus Innsbruck* (Innsbruck, 1893), 1-20

J. Müller[3] id., 'Kritische Studien zu den Naturales Quaestiones Senecae', *SAWW* 130.3 (Wien, 1894)

Münscher K. Münscher, 'Bericht über die Seneca-Literatur aus den Jahren 1915-1921', *JAW* 192.2 (1922), 109-214

F. Muller F. Muller, 'Ad Senecae Naturales Quaestiones observatiunculae', *Mnemosyne* 45 (1917), 319-37

Muretus see under Editions

OLD *Oxford Latin Dictionary*, ed. P. G. W. Glare (Oxford, 1982)

Oltramare see under Editions

Oltramare[1] P. Oltramare, 'Le codex Genevensis des Questions Naturelles de Sénèque', *RPhil* 45 (1921), 5-44

Otto A. Otto, *Die Sprichwörter und sprichwörtlichen Redensarten der Römer* (Leipzig, 1890)

Parroni[1] P. Parroni, 'Tre congetture alle *Naturales Quaestiones* di Seneca', *Dicti Studiosus: scritti di filologia offerti a Scevola Mariotti dai suoi allievi* (Urbino, 1990), 101-15

Parroni[2] id., 'Sul contributo del Genevensis lat. 77 al testo delle *Naturales Quaestiones* di Seneca', *RFIC* 120 (1992), 165-73

Parroni[3] id., 'Problemi testuali nelle Naturales quaestiones di Seneca', *Lingue tecniche del greco e del latino, Atti del I° Seminario internazionale sulla letteratura scientifica e tecnica greca e latina*, ed. S. Sconocchia, L. Toneatto (Trieste, 1993), 43-7

Parroni[4] id., 'Osservazioni sul testo delle "Naturales Quaestiones" di Seneca', *Scritti classici e cristiani offerti a Francesco Corsaro* (Catania, 1994), 537-48

Pincianus *Fernandi Pinciani ... in omnia L. Annei Senecae philosophi scripta ex vetustissimorum exemplarium* collatione castigationes utilissimae (Venezia, 1536)

RE — *Real-Encyclopädie der classischen Altertumswissenschaft*, ed. Pauly, Wissowa, et al. (Stuttgart, 1894-)

Rehm[1] — A. Rehm, 'Anlage und Buchfolge von Senecas Naturales Quaestiones', *Philologus* 66 (1907), 374-95

Rehm[2] — id., 'Das siebente Buch der Naturales Quaestiones des Seneca und die Kometentheorie des Poseidonios', *SBMünchen* 1921.1 (München, 1922)

Reinhardt — K. Reinhardt, *Poseidonios* (München, 1921)

Reynolds — L. D. Reynolds, *The Medieval Tradition of Seneca's Letters* (Oxford, 1965)

Rossbach[1] — O. Rossbach, 'XIMAIPA—AIΓEC', *NJPhP* 55 (1885), 35-7

Rossbach[2] — id., *De Senecae philosophi librorum recensione et emendatione* (Breslau, 1888; Hildesheim, 1969)

Rossbach[3] — id., review of Schultess[2], *DL* 10 (1889), 940-2

Rossbach[4] — id., review of Gercke[1], *BPhW* 17 (1897), 105-10

Rossbach[5] — id., ap. Gercke

Rossbach[6] — id., review of Gercke, *BPhW* 27 (1907), 1478-90

Ruhkopf — see under Editions

Schönberger and Schönberger — O. & E. Schönberger, *Lucius Annaeus Seneca, Naturwissenschaftliche Untersuchungen in acht Büchern, eingeleitet, übersetzt und erläutert* (Würzburg, 1990)

Schottus — A. Schottus, *Observationum humanarum libri V* (Hanover, 1615), 255-60

Schultess[1] — F. H. A. Schultess, *De L. Annaei Senecae quaestionibus naturalibus et epistulis* (Diss. Bonn, 1872)

Schultess[2] — id., *Annaeana Studia, Grat. Schr. des Hamburger Johanneums* (Hamburg, 1888)

Setaioli — A. Setaioli, 'Elementi di sermo cotidianus nella lingua di Seneca prosatore', *SIFC* 52 (1980), 5-47

Shackleton Bailey — see Bailey

Skutsch — F. Skutsch, ap. Gercke

Thulin[1] — C. O. Thulin, 'Fulgur, fulmen und Wortfamilie', *ALL* 14 (1906), 369-91, 509-14

Thulin[2] — id., *Die Etruskische Disciplin*, I, *Göteborgs Högskolas Årsskrift* 11.5 (1905)

TLL — *Thesaurus Linguae Latinae* (Leipzig, 1900-)

Vassis — S. Vassis, 'L. Annaei Senecae Quaest. nat. libri', *EE* 1907-8, 101-9

Vottero — see under Editions

Vottero[1] — D. Vottero, 'Problemi di critica del testo nelle "Naturales Quaestiones". I. L'ordinamento dei libri', *AAT* 107 (1973), 249-69

Vottero[2] — id., 'La grafia dei termini d'origine greca nelle opere filosofiche di Seneca', *AAT* 108 (1974), 311-39

Vottero[3] — id., 'Emendamenti al testo delle Naturales Quaestiones di Seneca', I, II, *AAT* 114 (1980), 323-46, 347-67

Vottero[4] — id., 'Note sulla lingua e lo stile delle "Naturales quaestiones" di Seneca', *AAT* 119 (1985), 61-86

Walter[1] — F. Walter, 'Zu lateinischen Schriftstellern', *Philologus* 80 (1925), 437-53

Walter[2] — id., 'Zu Seneca, Tacitus, Vergil', *PhW* 46 (1926), 189-92

Walter[3] — id., 'Zu Curtius, Seneca, Tacitus', *PhW* 47 (1927), 1565-8

Walter[4] — id., 'Zu lateinischen Schriftstellern', *Philologus* 83 (1928), 316-30

Walter[5] — id., 'Zu Curtius, Seneca, Tacitus', *Bayerische Blätter für das Gymnasial-Schulwesen* 66 (1930), 254-7

Walter[6] — id., 'Zu lateinischen Schriftstellern, II', *WS* 48 (1930), 182-7

Walter[7] — id., 'Zu Seneca, Festus, Porphyrio', *PhW* 51 (1931), 285-7

Warmington — E. H. Warmington, ap. Corcoran

Watt	W. S. Watt, 'Notes on Seneca, *Epistulae* and *Naturales quaestiones*', *CQ* 44 (1994), 185-98
Weinstock	S. Weinstock, 'Libri fulgurales', *PBSR* 19 (1951), 122-53
Winterbottom	M. Winterbottom, review of Corcoran, *CR* 26 (1976), 46-8
Wistrand	E. Wistrand, *Miscellanea Propertiana* (*Studia Graeca et Latina Gothoburgensia*, 38, Göteborg, 1977)

INTRODUCTION

THE MANUSCRIPT TRADITION

I have described the manuscript tradition in the preface to my edition, and elsewhere (especially Hine[3], Hine[9]). Here I simply give a list of sigla and a stemma. From the bifid stemma it will be clear that one often has to decide between the readings of ζ and Ψ, a frequent preoccupation of my notes.

Principal Manuscripts

A	Leiden, Bibliotheek der Rijksuniversiteit, Voss. Lat. O.55 (s. XII[1])
B	Bamberg, Staatsbibliothek, Class. 1 (M.IV.16) (s. XII)
F	Oxford, Merton College, 250 (s. XII)
γ	the text of fragments in the Florilegium Duacense (γ is the consensus of five manuscripts, see Hine[3])
H	Paris, Bibliothèque Nationale, lat. 8624 (s. XII[2])
L^2	corrector of L, Leiden, Bibliotheek der Rijksuniversiteit, Voss. Lat. F.69 (s. XII)
P	Paris, Bibliothèque Nationale, lat. 6628 (s. XII[2])
R	El Escorial, Real Biblioteca, O.III.2 (s. XIII)[1]
U	München, Bayerische Staatsbibliothek, Clm 11049 (s. XV)
V	Vatican City, Biblioteca Apostolica Vaticana, Pal. lat. 1579 (s. XII[2])
W	Venezia, Biblioteca Marciana, Lat. Z.268 (1548) (s. XIV)
Y	München, Bayerische Staatsbibliothek, Clm 18961 (s. IX[2]; excerpts)
Z	Génève, Bibliothèque Publique et Universitaire, lat. 77 (s. XII[m]).

[1] R is a composite manuscript whose text derives partly from α (from 1.pr.1 to 1.13.1, from 2.53.2 to 3.pr.18, and from 7.15.2 to 7.32.4), partly from ζ (from 1.13.2 to 2.53.2), and partly from π (from 3.pr.18 to 7.15.2); see Hine[2].

ζ	consensus of RZ (from 1.13.2 to 2.53.2)
δ	consensus of ABV[2]
α	consensus of δR (from 1.pr.1 to 1.13.1, from 2.53.2 to 3.pr.18, and from 7.15.2 to 7.32.4)
θ	consensus of FH
ρ	consensus of PR (from 3.pr.18 to 7.15.2)
υ	consensus of UW
π	consensus of ρυ (or Pυ where R does not derive from ρ)
Ψ	consensus of αθπ (or δθπ where R does not derive from α, or θπ where δ is missing)
Ω	consensus of ζΨ or (ZΨ where R does not derive from ζ)

Stemma

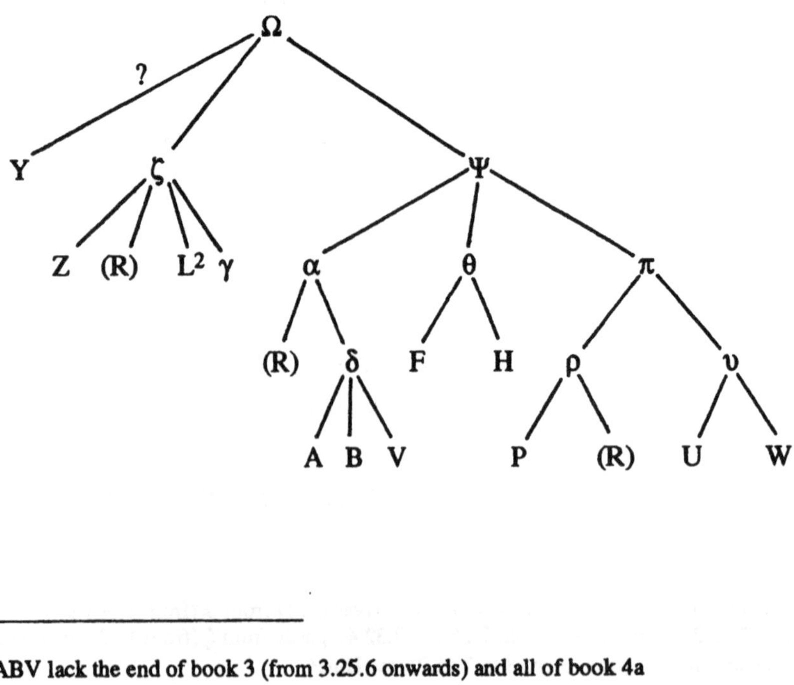

2 ABV lack the end of book 3 (from 3.25.6 onwards) and all of book 4a

Other manuscripts

The following manuscripts are occasionally cited:

C	Heiligenkreuz, Bibliothek des Zisterzienserstifts, 213-III (saec. XII^2)
Cam.	Cambrai, Bibliothèque Municipale, 939 (838) (A. D. 1379)
D	Dublin, Trinity College, 514 (saec. XIII)
E	Berlin, Staatsbibliothek Preussischer Kulturbesitz, lat. oct. 9 (saec. XIII)
f	El Escorial, Real Biblioteca, N.III.16 (saec. XIII)
G	Montpellier, Bibliothèque de la Faculté de Médicine, 116 (saec. XII^2)
J	Oxford, St. John's College, 36 (saec. XII/XIII)
k	Trier, Bibliothek des Priesterseminars, 66 (R.IV.2) (saec. XII/XIII)
K	Cambrai, Bibliothèque Municipale, 555 (513)-I (saec. XII^2)
L	Leiden, Bibliotheek der Rijksuniversiteit, Voss. Lat. F.69 (saec. XII^2)
O	—, B. P. L. 199 (saec. XIII^2)
S	Montpellier, Bibliothèque de la Faculté de Médicine, 445 (saec. XII/XIII)
T	Paris, Bibliothèque Nationale, lat. 6386 (saec. XIII)
X	Exeter, Cathedral Library, 3549B (saec. XIII)

ς	a reading found in one or more 14th or 15th century manuscripts (apart from UW)

A^1, B^1, etc. = A, B, etc., before correction

A^c, B^c, etc. = A, B, etc., corrected either by the first or by a later hand

A^2, B^2, etc. = A, B, etc., corrected by a later hand.

1

Notes on *Naturales Quaestiones* Book 1

1.pr.2 denique tantum inter duas interest quantum inter deum et hominem.

tantum Z, *post* duas *add.* L^2: *om.* Ψ

Earlier editions all had *tantum*, but Gercke and more recent editors omit it, without comment. Prima facie, ZL^2's reading should be accepted; and *tm̄* could easily have been omitted before *inter*; yet *tantum* would be a natural supplement, so their reading may be conjectural. But the omission of *tantum* is harsh and not, I think, properly paralleled. H-Sz 591 discuss some related phenomena, but we should ignore cases where (a) a comparative is involved, or (b) the correlative is omitted from the second clause, not from the first (cf. 7.14.4 *nam quo plures fuerint, saepius †in aliquas incident†*). Comparable, perhaps, is *non ... quam* for *non tam ... quam*: e.g. Plaut. *Rud.* 943 *non edepol piscis expeto quam tui sermonis sum indigens*; Liu. 24.23.9 *sed quod uolebant non quam maturato opus erat nauiter expediebant* (that is made easier because the *quam* comes first); Tac. *Ann.* 3.8.1 *quem haud fratris interitu trucem quam remoto aemulo aequiorem sibi sperabat*; but here we have no negative.

The passage needs to be considered with 7.2.1: ... *num cometae condicionis eiusdem* [*eiusdem* Z: om. Ψ] *sint cuius superiora* (*<eius> condicionis* Fortunatus). Elsewhere in Seneca the nearest parallel I have noticed is *Dial.* 10.18.3 *in officio amorem consequeris in quo odium uitare difficile est* (*in <illo> officio* Gertz, *in <eo> officio* mg. ed. de Juges); but that is easier. Probably the omission of *tantum* here and *eiusdem* at 7.2.1 is conceivable, but it is harsh, and I prefer to follow Z, as does Parroni[3], who notes here the chiastic parallel with the opening sentence of the book, *Quantum ... interest ... tantum ...*

1.pr.4 nisi ad haec admitterer, non fuerat operae pretium nasci.

operae pretium ZL[2]: *om.* Ψ

Before Z's reading was known there were various supplements on offer: *profuerat* D[2]T; <*opus*> *fuerat* Madvig[2]; *fuerat* <*tanti*> Cornelissen; <*tanti*> *fuerat* Leo, Rossbach[5] (comparing *Ep.* 76.6, *Dial.* 9.8.7), who are followed by Gercke and more recent editors; <*nauci*> *fuerat* Walter[7]. Again there is a question whether ZL[2] preserve a genuine reading, or give a conjectural supplement. Axelson[1] took the latter view, and proposed *iuuerat*. For *operae pretium* cf. *Dial.* 12.10.8 *cuius exitum nosse operae pretium est* (cf. in general *TLL* 9.2.667.23-71, esp. 32-48). Axelson's *iuuerat nasci* gives a better rhythm, but ZL[2]'s rhythm is quite acceptable, and there is no sufficient reason to reject their reading, as Parroni[3] says. (Compare 6.31.1: ... *quassa quatiebat, quibus ad cadendum male stantibus non erat impelli sed agitari* (*stantibus* <*opus*> Madvig[2]).)

1.pr.4 ut hoc corpus causarium ac fluidum, periturumque nisi subinde suppletur, farcirem ...?

suppletur Z: impletur RθPW: impleatur δ: impleretur U

impletur gives inferior rhythm, and inferior sense: for the body needs to be continually replenished, but continual *filling* sounds excessive, like *farcire* following. Cf. 3.4 *quid est quod illam* (sc. *terram*) *aut sic impleuerit ut praebere tantum ex recondito possit, aut subinde sic suppleat?*, 3.14.3.

1.pr.5 quamdiu cum adfectibus conluctamur, quid magnifici facimus? etiamsi superiores sumus, portenta uincimus: quid est, cur suspiciamus nosmet ipsos, quia dissimiles deterrimis sumus?

... facimus, etiamsi superiores sumus? portenta <sumus, nisi> uincimus *Shackleton Bailey*

Gercke's punctuation is given above. Shackleton Bailey rightly objects that overcoming *portenta* is normally a considerable achievement (though the context may show otherwise, as at Cic. *Tusc.* 1.11 *quid negotii est haec poetarum et pictorum portenta conuincere?*), hence his punctuation and supplement. But his supplement suggests that we *are* something if we do overcome our passions, whereas the whole thrust of the passage is that even overcoming our passions does not get us very far. So maybe punctuate as follows: *quamdiu cum adfectibus conluctamur, quid magnifici facimus, etiamsi superiores sumus? portenta uincimus: quid est, etc.* or *portenta uincimus? quid est, etc.*; with either punctuation *portenta uincimus* serves as a protasis, *quid est, etc.*, as an apodosis. Here *portenta* perhaps refers to vicious people: cf. 1.16.3 Hostius Quadra is called *portentum illud*; *TLL* 10.2.19.76-20.13. Otherwise to the vices themselves: cf. *Dial.* 5.15.2 *tam truci portento* (a cannibal feast); *Ep.* 92.9 *at isti sapientiam ex quibus composuere portentis?*, 110.12 *alia portenta luxuriae*; Plin. *Nat.* 37.50 *Domitius Nero in ceteris uitae suae portentis ...* (*TLL* 10.2.19.43-75).

1.pr.10 ... equitem modo extrema cingentem, modo ulteriora explorantem, modo a lateribus adfusum ...

extrema cingentem modo ZL^2: *om.* Ψ

There should be no doubt that *modo extrema cingentem* is genuine, and omitted by parablepsy in Ψ; *extrema cingentem* is protecting the rear (cf. e.g. Liu. 6.32.11, 21.34.7 for *extrema agminis*), *ulteriora explorantem* refers to advancing in front of the column, the last phrase to protecting the flanks.

1.pr.10 quid illis et uobis interest nisi exigui mensura corpusculi?

illis et uobis VHP: illis et nobis RABFυ: inter illas et uos Z

There is no parallel for the dative with *interesse* (Apul. *Met.* 11.27.3 is ambiguous, see *TLL* 7.1.2284.35-7), but it seems perfectly intelligible as an ordinary dative of person concerned, so Z may offer a normalisation.

1.1.6 tunc ignes tenuissimi iter exile designant et caelo producunt.

After *iter exile designant*, the clause *et caelo producunt* adds nothing material to the meaning, and the style is bland. The clausula could be paralleled, but it is perhaps surprising S. did not write *caeloque producunt*. So I wonder if the clause is a gloss on *iter exile designant*, an unusual and striking expression. Alternatively one could emend *producunt*: Professor Nisbet suggests *prodeunt*.

Gemoll[2] proposed *<in> caelo*, being justifiably worried by the bare ablative; but cf. perhaps 6.12.2 *per superiores dies caelo aër stetit*; Sall. *Iug.* 17.5 *caelo terraque*; it is mostly poetic, of course, see *TLL* 3.86.80-88.12 passim; cf. Cic. *Fin.* 5.9 *natura sic ab iis inuestigata est ut nulla pars caelo, mari, terra (ut poetice loquar) praetermissa sit.*

1.1.9 illud enim stultissimum, existimare aut decidere stellas aut transilire aut aliquid illis auferri et abradi: nam si hoc fuisset, etiam defuissent.

fuisset Ψ: esset Z etiam ZRθπ: iam δ: et Cam.E defuissent] defecissent H[1], *ut coni. Madvig*[2]: redissent *uel* reduces essent *Gercke in app.*

The second half of the sentence is awkward. Madvig[2] objected to the tense of *defuissent,* saying we would need *deessent;* however, the pluperfect could mean 'they would already have disappeared'. *etiam* could mean 'they would actually have disappeared', but *iam* is much better sense ('they would already have disappeared before now'), and duplication of the ending of *fuisset* before *iam* was very easy. In the protasis, the jingle *fuisset/defuissent* is maybe possible, with the play on simple and compound verb, but Z's *esset* is preferable stylistically (avoiding the jingle) and in sense: 'if this were true, they would already have disappeared'. For this use of *esse* cf. *Ep.* 66.25 *aut si hoc est, magis diliges ...,* 82.18 *si hoc est, perit gloria* (and maybe *Nat.* 6.14.2, but there is a lacuna before *nisi hoc est,* so we cannot be sure).

1.2.8 cum ab una parte cesserunt (sc. coronae), illinc uentus est unde finduntur.

ab una parte *Pincianus*: ob unam partem ZRθPW[1]: ad unam partem δU: in unam partem f

Seneca is talking about a break occurring at one point of a halo, as opposed to the whole thing dissolving (*cum dilapsae sunt aequaliter et in semet ipsae euanuerunt*) or breaks occurring at several points (*si ruptae pluribus locis sunt*). The archetype must have had *ob unam partem,* which scarcely means anything. δU's *ad unam partem* would convey the wrong sense, meaning that the whole halo moved in a particular direction; and in any case Seneca would have written *in unam partem* to convey that sense: *ad ... partem* occurs only three times in his works, none comparable (*Ben.* 2.18.1, 5.11.1, *Ep.* 8.9), whereas *in ... partem* occurs some dozens of times, and *in unam partem* at *Nat.* 5.1.1 (bis), 5.1.3 (ter), 5.1.4, 5.8.1, 5.16.5, 7.6.1. Pincianus' conjecture not only gives the right sense but improves the clausula. Cf. particularly 1.2.5 *nonnumquam ab aliqua parte rumpuntur,* 1.2.9 *si ab una parte intercisa est.* (Also 1.8.1, 1.13.3, 1.14.5, al.)

1.2.9 ex his quae iam posui

posui Z: exposui θπ: proposui α

The simple verb *pono* can mean 'state, say', and is just as apposite here as *expono* or *propono*; cf. *OLD* s.v. *pono* 18a; e.g. Cic. *de Orat.* 2.271 *multa rettulit, ex quibus a me exempli causa non nulla ponuntur, Brut.* 219 *ut ne in scripto quidem meminisset quid paulo ante posuisset*; Gell. 12.2.11 *et cum reprehendisset uersus quos supra de Cetego posuimus...* In both the Cicero passages the simple verb gives a preferable rhythm to either compound. No Senecan example of *pono* is quite parallel. (For *proposui* cf. e.g. *Nat.* 1.2.7, *Dial.* 1.3.2, for *exposui*, *Dial.* 6.18.1.) But, even though the simple verb has no rhythmical advantage, Z may have the true reading for which θπ and α substitute more familiar compounds. Cf. 2.9.3 *pressa* ζ, *depressa* Ψ; 2.34.2 *seruata*] *obseruata* δ; 3.18.6 *sequitur*] *prosequitur* δ (where *sequitur* should be accepted). See on simple verbs for compounds, H-Sz 298-300, Axelson[1] 88-90, Axelson[2] 114, 202; Reynolds, 126; Hine[6] on *Nat.* 2, locc. citt.; below on 6.1.2.

1.3.4 poterat enim uerum uideri si arcus duos tantum haberet colores, si ex lumine umbraque constaret; nunc ...

nunc Z: sed nunc Ψ

After an unreal supposition, to say 'but as things really are', Seneca elsewhere uses *nunc* (*Nat.* 1.5.10, *Dial.* 12.5.2, 12.15.3) or *at nunc* (*Dial.* 6.20.4, *Nat.* 4a.2.25; and after an open condition at *Ep.* 19.7), but never *sed nunc* in this sense (contrast the temporal uses at *Nat.* 4a.2.17, *Ep.*

30.11, 68.12, *HF* 408, *Ag.* 241; *at nunc* in temporal sense at *Nat.* 3.18.4, *Ep.* 86.5, 86.8). So *nunc* is here preferable (and was easily augmented in Ψ).

1.3.4 nam commissura decipit, usque eo mira arte naturae; quod a simillimo coepit, in dissimillimo desinit.

More recent editors punctuate as above, and translators apparently take *usque eo* with *mira* (e.g. Vottero 'tanto mirabile è l'arte della natura'), but I know no parallel for *usque eo* with an adjective. Rather punctuate thus: *nam commissura decipit: usque eo, mira arte naturae, quod a simillimo coepit in dissimillimo desinit.* Normally one finds *usque eo ... ut*, but here the consequence is given first (*nam commissura decipit*), then the explanation. Cf. the passage of Ovid just quoted by S., *Met.* 6.67 *usque adeo quod tangit idem est, tamen ultima distant.*

1.3.6 itaque piscinam ingentis magnitudinis insertis parietibus diuide: totidem illa habebit imagines solis quot lacus habuerit; relinque illam sic ut est: diffusa semel tibi imaginem reddet.

Punctuation before *diffusa* is undesirable, for it spoils the emphasis of the next clause, which is on *semel.* On the other hand, *diffusa* adds nothing material to the sense of the previous clause—*sicut est* is enough—so perhaps it is a gloss. (For this sense of *diffundo* cf. *OLD* s.v. 1a.)

Probably *sicut* should be printed as one word, for consistency (as several manuscripts have it). (*sic ut* at *Ben.* 5.23.2, *Ep.* 109.11, *Nat.* 1.5.9, is different, for the *ut* is final.)

1.3.8 ergo cum multa in pluuia stillicidia sint, totidem specula sunt.

in pluuia Z: *om.* Ψ

A gloss is possible, but so is omission (*multa in pluuia*), and if *cum ... sint* is right, the causal clause is in favour of *in pluuia* (otherwise it ought to be temporal). The corresponding sentence of Aristotle *Mete.* 3.4 runs (373b15-7): ἕκαστον γὰρ τῶν μορίων ἐξ ὧν γίγνεται συνισταμένων ἡ ψακὰς ἔνοπτρον ἀναγκαῖον εἶναι μᾶλλον τῆς ἀχλύος. Parroni[3] compares 1.3.5 *singula stillicidia pluuiae cadentis*, 1.3.6 *stillicidia ... quae imber cadens defert*, 1.3.10, for specification of the source of the *stillicidium*.

1.3.10 ad ipsum solem reuertere: hunc, quem toto terrarum orbe maiorem probat ratio, acies nostra sic contraxit ut sapientes uiri pedalem esse contenderent; quem uelocissimum omnium scimus, nemo nostrum moueri uidet, nec ire crederemus nisi appareret isse.

In the second half of the sentence the reference is to Saturn, the outermost of the planets known in antiquity, and the nearest in speed to the celestial sphere, which is mentioned in the next sentence. Gemoll[2] suggested adding *sidus* after *scimus*, which clarifies the sense attractively; but better rhythm is obtained by adding *siderum* before *scimus* (where it was easily omitted after *omnium*).

1.4.1 <Et>iamnunc illud accedit, quod aeque manifestum est ...

suppl. Haase

Oltramare says that *etiamnunc* is redundant with *accedit*; but cf. 6.18.6 *etiamnunc et illud accedit*; *Ben.* 3.8.2 *etiamnunc adice*; *Ep.* 92.12 *etiamnunc adiciam*. So the question is whether the paradosis *iamnunc* or *iam nunc* will do, in what must be a transitional sense, not purely temporal. The transitional use, according to *TLL* 7.1.122.63-5, is found in Tertullian (*test. anim.* 4.1 *iam nunc quod ad ... sententiam tuam ... spectet ..., adfirmamus ...*; *adv. Marc.* 1.5.3, *adv. Hermog.* 19.2, *Spect.* 10.10). For *iam nunc* in the temporal sense see *TLL* 7.1.114.4-42, on *nunc iam* ib. 42-64. Seneca nowhere else has *iam nunc*, and *nunc iam* only at *Ag.* 371 and *Med.* 364. So Haase's conjecture certainly brings the passage into line with Seneca's usage elsewhere, but it is feasible that he anticipated a usage next found in Tertullian.

1.5.3 qui montium summa conscendunt, spectant nubem, nec ...

inspectant *Haupt[1]*: despectant *Madvig[2]*: superstant *Schultess[2]*

Haupt: '... nam ut spectemus nubem non opus est ut summos montes conscendamus.' But the context implies that the mountaineers will have a closer look (*OLD* s. *specto* 6). As for Madvig and Schultess, why should looking down on clouds be more instructive than looking at them closely from below?

1.5.8 quid enim simile speculo habent nubes? cum illa non perluceant, hae transmittant lucem.

simile speculo (speculis δ) habent Ψ: habent speculo simile Z

Does *illa* require *speculis* in the previous sentence? Such changes are documented by K-S 1.63-4, though they give no examples from Seneca; but here the plural *nubes* encourages the change to plural *illa*. Z's rhythm is preferable, avoiding the elision of *speculo*.

1.5.9 nam quemadmodum speculo opponendum est id cuius in se imaginem transferat, sic, ut nubes infici possint, ita sol ad hoc apte ponendus est; non enim idem facit undecumque effulsit, et ad hoc opus est radiorum idoneus ictus.

undecumque Z: si undique Ψ   ictus] situs *Axelson*[1]: tractus *Watt*

undecumque is the required sense, not *si undique*. The last clause arouses several worries:

(i) *et* is odd: should it be *sed* (Bongarsius)—but what is the contrast? or should we punctuate ... *effulsit: et ad hoc* ..., 'for this too [for the rainbow, as well as for mirrors] a suitable ... is needed'? This punctuation gets round a second problem, that *ad hoc* is unnecessary (and also occurs in the previous sentence).

(ii) *opus est* with the nominative is unparalleled in Seneca (this is the only Senecan example in *TLL* 9.2.860.38; of course *quid opus est* and the like are not really parallel, cf. *Ben.* 2.11.3); but this is no strong objection: in Livy too only one case of the nominative is forthcoming (*TLL* 9.2.860.36).

(iii) The heroic clausula is very rare in S.

Is the whole clause, *et - ictus*, maybe a gloss, as Ruhkopf thought? But, with *undecumque*, deletion leaves an unimpressive clausula. So do we accept Axelson's conjecture? But *situs* seems too static a word for the sun's rays. Watt suggests *tractus*, 'direction', but in this sense it seems to be used only of concrete physical objects. Maybe we should read ... *radiorum idoneus angulus (a(n)g(u)lus → ictus).* For the geometrical sense of *angulus* cf. 5.17.3 *rectis angulis, OLD* s.v. 1.

1.5.10 ... qui speculari ratione ...

speculari Ψ: speculi Z

specularis is found elsewhere only as an adjective in the phrase *lapis specularis* or as a noun, *speculare,* so Z should be followed.

1.5.14 sunt (specula) quae dextras facies ostendant, sunt quae sinistras, sunt quae detorqueant et uertant

The clausula would be improved by supplying *<in>uertant*; cf. 1.4.3 *inuersi tibi uidebuntur et propiores a te quam a speculo*; *Ep.* 121.8. The prefix *i(n)* was vulnerable to omission before *u.*

1.6.3 dices enim fortasse unde sit illi color uarius; unde talis figura sit, non dices, nisi ...

figura sit BVHπ: figura ZF: sit figura RA

See K-S 1.12 for ellipse of the verb in indirect questions; there is no example there from S., but here the preceding *unde sit* ... makes the ellipse easy, and the rhythm is better than with *figura sit* (-ūră nōn dīcēs; the order *sit figura* is only weakly attested).

1.6.3 nullum autem (exemplar) quam solis est, a quo ...

<aliud> quam δ

Haase was first to omit *aliud,* followed by subsequent editors; but is it possible to do so? *quam* is used for *magis* or *potius quam,* cf. H-Sz 593-4, *OLD* s. v. *quam* 9 (e.g. S. *Ben.* 1.9.5 *certant ... suam (paupertatem)* [*magis* add. Gertz] *quam ullum aliud uereri malum*). But here the sense required is *aliud* rather than *magis* or *potius.* So we should probably accept δ's reading: *aliud* abbreviated was vulnerable after *autem* abbreviated. For *nullum aliud quam* in S. cf. *Ep.* 76.11 *nullum aliud est hominis quam ratio*; *Nat.* 2.9.2. (Before Livy *nihil aliud nisi* was the norm; see H-Sz 595.)

1.6.6 ... est aliquid et aquae, quam, etiamsi nondum habet, iam parat, id est iam eius natura est in quam ex sua uertatur.

uertatur (uersatur R) αθPU: uertitur ZW[1]

A *locus nondum sanatus* that has produced various attempts to translate the paradosis:

'that is, there exists already the property of the water into which the cloud may be changed from its own property' (Corcoran);
'autrement dit, quelque chose qui a déjà cette manière d'être que le nuage prendra en abandonnant sa nature propre' (Oltramare);
'es decir, existe ya la naturaleza que adoptará abandonando la suya propia' (Codoñer Merino);

and also various conjectures have been proposed for *iam, etc.*:

iam eius naturae est in quam ex sua uertitur Gemoll[2] (def. Axelson[1]);
iam eius natura est <quae> in <a>quam ex sua uertatur F. Muller 324-5;
essentiam eius nactura est in quam ex sua uertatur Garrod[1] (deleting *id est essentiam - uertatur* as a gloss on the previous clause);
iam eius naturam nactura est in quam ex sua uertatur Münscher 165;
similis eius natura est in quam ex sua uertatur Vottero[3];
iam eius<modi> natura est in quam ex sua uertatur Watt *per litteras*

Rhythm recommends *uertitur*. The sense we need could be achieved by: *id est iam <paene> eius naturae est in quam ex sua uertitur*: 'that is, it is now virtually of that nature (*OLD* s.v. *natura* 8; genitive of description) into which it is changing from its own nature'; or, more freely, 'that is, it has already virtually reached the state into which it is changing from its present state'. The sentence is possibly a gloss, as Garrod thought.

1.11.2 in propinquo fere a sole uisuntur

I cannot give another example of *propinquus* plus *ab* (though H-Sz 104 refer to it as normal except with names of towns and countries); it is analogous to *prope* and *proximus.*

1.11.3 sunt autem imagines solis in nube spissa et uicina in modum speculi.

uicina et spissa Z

With Ψ's word order, *spissa et uicina in modum speculi,* Garrod[1] remarks that *uicina* seems nonsensical, and he suggests changing *uicina* to *incurua,* which is accepted by Vottero in his edition. However, Z's word order is easier, for *spissa* is then next to *in modum speculi* and can be taken closely with it. For thickness lending reflective properties, cf. 1.3.7 *ergo etiam ab aëre spisso uisus noster in nos redit,* 1.12.2; 1.13.2. As for *uicina,* the point is that, in order to form a parhelion, the cloud must be in the same part of the sky as the sun (in contrast to rainbows, where the cloud is in a different part of the sky). Cf. later in 1.11.3: *sequitur enim illum [sc. solem nubes] nec umquam longius relinquitur quam fuit cum apparuit.*

1.13.1 quid enim impedit quominus tot sint (parhelia) quot nubes fuerint aptae ad exhibendam solis effigiem?

fuerunt ZRAVθπ

fuerunt is the paradosis, and was in all editions prior to Fickert's. The indicative is expected (cf. e.g. *Ben.* 5.1.5 *per te uero non est mora, quo minus beneficia, qui acceperunt, ultro repetant, Dial.* 1.5.5 *quidni nullam moram faciam quominus recipiatis quod dedistis?*, 7.18.2, *Ep.* 26.10 (but contrast *Nat.* 6.7.5 *nec enim uideo quid prohibeat aut obstet, quominus habeat ... mare, quod illic quoque tantundem loci teneat..., Ep.* 94.51 *impedietur quominus ad illud perueniat quo possit se esse contentus*)). The perfect tense, however, is unexpected, but can be seen as part of S.'s habitual variation of tense (just as in 1.12.1 he has *cum uterque orbis sub eodem libramento stetit* and 1.12.2 *cum sic coactus aër et limpidus constitit* ...). If change is needed, *sunt* would be simple (misread as *fu(er)u(n)t*).

1.13.2 debent autem hae nubes, quae hoc praestant, densae esse, leues, splendidae, planae, †naturae solis†.

planae L^2H: plane Zαπ: plene F   natur(a)e solis Ω: naturae solidae Cam.DTX, *ut coni. Larisch*[3]

The vulgate was *plane naturae solis*, but Madvig[2] objects to this on grounds of Latinity ('Neque Latine dicitur esse naturae solis, ut sit, eiusdem, atque sol sit, naturae (etsi dicitur esse figura tauri)') and sense, for it could occur to nobody to attribute the nature of the sun to clouds. F. Muller unconvincingly defends the paradosis, comparing the problematic text of 1.6.6 (discussed above). Conjectures are numerous:

plane saturae solis Madvig[2];
plane naturae solidae Cam.DTX, Larisch[3];
planae, naturae solis <similes> Leo (cf. 1.11.3 *quidam parhelion ita definiunt: nubes rotunda et splendida similisque soli*);
plane maturae soli Rossbach[6] 1481;
planae, imaginaturae (or *nacturae*) *soles* (or *ut natura est solis*) Garrod[1];
planae, <uice fu>ncturae solis Oltramare;
planae, naturae solis <simillimae> Alexander;
planae, naturae similes solis Codoñer Merino;
planae, aduersae solis Codoñer Merino in app.

1. Madvig's *saturae solis* is doubtful because of the genitive with *satur* (found in Ter. *Ad.* 765 and Hor. *Ep.* 1.7.35, according to *OLD*, but the ablative is commoner, and found in S. at *Ep.* 18.7, *Thy.* 955), and because of the sense (would S. describe clouds thus?).

2. *naturae solidae* is implausible: for S. never says clouds are solid (contrast the comparative at 1.2.11 *in materia solidiore*), and if it just means 'dense' it repeats *densae* just before.

3. *naturae solis <similes>* or *<simillimae>* is implausible, for why not just say *soli similes*? S. might say the clouds resemble the form of the sun, but why the nature of the sun?

4. *maturae soli* has nothing to recommend it, nor do Garrod's conjectures.

5. Oltramare's proposal is bizarre: the reflection conceivably plays the role of the sun, but the cloud does not.

6. Codoñer Merino's first suggestion is subject to objection 3 above; the second, *aduersae solis,* is a novel approach, but what does *aduersae* mean? The clouds must be close to the sun, not facing (opposite to?) it. And does adjectival *aduersus* ever take the genitive? (With *ex aduerso,* e.g. *cuius ex aduerso,* cf. *TLL* 1.868.54-63, 74-77; with neuter plural *aduersa, TLL* 1.872.69-82; but what we have here is not comparable.)

Another possibility is *uicinae soli* (not *uicinae solis,* because *uicinus* always takes the dative in S.); cf. c. 11.3 *in nube uicina et spissa,* on which see above. Professor Nisbet suggests *naturae* (gen. of description) *soli <simillimae>.*

1.14.1 horum plura genera conspiciuntur: sunt putei, cum uelut corona cingente introrsus ingens caeli recessus est similis effossae in orbem specus ...

putei *Gercke, Rossbach*[5]: ut ei ζθπ: *om.* δ: βόθυνοι *Casaubon ex cod. suo, teste Grutero*[2]: cute *uel* ceuthe *Garrod*[1]: cute *uel* κύτη *F. Muller*: bothyni *Vottero*[2] *319-20.*

Seneca is probably talking about auroral displays, here describing one which is like a bright ring with, as it were, a receding cave inside it. Such things are called βόθυνοι at Arist. *Mete.* 342a36 and *Mu.* 392b4, and the Latin *bothyni* is found at Apul. *Mun.* 3, 16. Casaubon championed βόθυνοι here, which he found in a manuscript, and Vottero preferred *bothyni.* But this is rather distant from the paradosis *ut ei.* Hence Gercke and Rossbach suggested *putei* as a Latin equivalent of βόθυνοι; the word is not found elsewhere in this sense, but references to such phenomena are rare anyway. Garrod[1] objects that, since the other two categories S. goes on to mention are Greek loan words, *pithiae* and *chasmata,* we need the same here; Vottero argues similarly. But the argument is weak: cf. 1.15.4, where Latin *trabes* occurs alongside Greek *pogoniae, cyparissiae, lampades* and *pithiae.* (Also, the word *chasma* seems at home in Latin: at *Nat.* 6.9.2 it refers to a fissure in the earth, a usage found in legal writers: Scaevola *dig.* 18.6.11; Ulp. *dig.* 30.1.47.6, etc., see *TLL* s.v.; and it is also used of celestial phenomena at Plin. *Nat.* 1.2.26, 2.96.) Garrod's and Muller's suggestions have little to commend them, for neither κύτος nor κεῦθος is found in the required sense in Greek writers (the nearest is Vett. Val. 4.11.11 ἀστέριον κύτος, the vault of the heavens). So on balance *putei* is best: it is near to the paradosis and satisfactory sense.

1.14.1 similis effossae in orbem specus

effoss(a)e Ψ: -ssa Z   specus θπ: specu ζ: spelunce δ

similis usually takes the dative in S., except for *ueri similis*, and *similis* with the genitive of personal pronouns. The only sure example of the genitive otherwise is *Dial.* 3.12.5 *pacem enim similem belli efficit* (cf. Oltramare[1] 31). The final *s* of *specus* could come from *sunt* at the start of the next sentence.

1.15.3 haec fulgores dicuntur

Schultess[1] 12 n. 2 proposes *fulgura*, because a genus of the species *fulgores* cannot be given the name *fulgores*; editors follow suit. Logically he is right, but the usage that S. records (*dicuntur*) may not fit logically with his own terminological scheme, so it is best to leave well alone.

1.15.3 ab his tacta nos dicimus icta sine fulmine, quae asperoplecta Graeci uocant.

tacta H^{c}π: acta ζ: tracta FH^{1}: tecta δ   <fulgurata id est> icta *Gercke*: <fulgurita, id est> icta *Castiglioni ap. Oltramare*: <siderosa id est> icta *Codoñer Merino*: <siderata id est> icta *Thulin[1] 382-3*: <fulgurita, id est fulgure> icta *Vottero[3]*   asperoplecta ζL^{2}θπ: asperoplectra AB: aspersoplectra V: astrapoplecta *Erasmus[2]*: ἀστραψόσκηπτα *uel* ἀστραψοπρηστῆρας *uel* ἀστραψόπληκτα *Fortunatus in comm.*: ἀστερόπληκτα *Jan 262*: ἀστραπόπληκτα *Vassis*

We cannot be absolutely certain what S. wrote here. The manuscripts point to *quae* ἀστερόπληκτα *Graeci uocant* (Jan), rather than ἀστραπόπληκτα (Vassis). On the other hand we might be led to expect ἀστρόπληκτα by ἀστροπλήξ in *CGL* 2.183.41 and τὰ ἀστρόπληκτα in Jo. Alch. 266.14, along with the analogy of ἀστρόπληγος (Gp. 5.36.1), ἀστρο–βλής, –βλητός, –βολησία, etc. (see LSJ). But ἀστεροβλὴς κεραυνός (*Orphica* fr. 32c5 K, cf. Weinstock 129) supports ἀστερόπληκτα, and see below on the etymology.

Some Latin word must have fallen out, for *icta sine fulmine* looks like an explanation of a technical term, rather than the Latin term itself that corresponds to the Greek. Suggestions are:

(1) Gercke's *fulgurata*: but transitive *fulgurare* is rarely found—see below on 2.21.2, where one should maybe alter to *fulgurita*;

(2) Castiglioni's *fulgurita*, which is better, because this is the well-attested religious term (see *TLL* and *OLD* s.v., including *Dial.* 5.23.6).

But could either of these be glossed by *icta sine fulmine*? This would presuppose a sharp distinction between *fulgur* and *fulmen*, and also presupposes Schultess's *fulgura* above. (Vottero[2] 326, and Vottero[3], sharpens the conjecture up with *<fulgurita id est fulgure> icta sine fulmine.*) But in ordinary language the *fulmen/fulgur* distinction was not so sharp, still less in religious language (cf. Hine[6] on *Nat.* 2.1.3, Paul. Fest. p. 92 *fulguritum, id quod est fulmine ictum ...*).

Therefore *siderata* (Thulin[1] 382-3, cf. Plin. *Nat.* 9.58 *silurus caniculae exortu sideratur,* 28.226 *sideratis urina ... prodesse dicitur,* and *sideratio* at Plin. 17.218, 222, 226; *CGL* 2.324.54 ἡμίξηρος *siderata sideratus*) or *siderosa* (Codoñer Merino ad loc., cf. *CGL* 2.183.41 *siderosus* ἀστροπλήξ, ἐπιληπτικός) should be considered. But with these supplements *icta sine fulmine* is a curious explanation, and maybe we should follow Thulin's observation that it is explaining ἀστερόπληκτα as if it were derived from α-privative + στεροπή + πλήγω. But if *icta sine fulmine* were Seneca's gloss on the Greek word it ought to follow it. Hence my conjecture (Hine[7]) that *icta sine fulmine* is a gloss on ἀστερόπληκτα, and has ousted either *siderata* or *siderosa*. The former is slightly preferable, given the Pliny parallels. So I would read: *ab his tacta nos dicimus <siderata> [icta sine fulmine], quae* ἀστερόπληκτα *Graeci uocant.*

1.16.1 Hostius fuit Quadra, obscenitatis in scaenam usque perductae.

scenam ZABHP: cenam RVFυ: canitiem *Shackleton Bailey*: senium *Watt* producte ς

in cenam is fairly well attested, but a feeble alternative, and not elucidated by what follows. *in scaenam* has been interpreted in three ways:

1. Hostius was so notorious that his behaviour was mentioned or enacted in stage performances (Gronovius[2] 252 'Videlicet adeo notae, ut impune histrionum et mimorum dictis traduceretur'), or else audiences applied certain dramatic verses to him (for this happening cf. Cic. *Att.* 2.19.3, Suet. *Tib.* 45, Cic. *Sest.* 118-23). But as Shackleton Bailey 450 says, '[i]f that had been what Seneca had in mind, he would have made himself intelligible.'

2. Hostius himself took part in obscene performances on stage (cf. 7.32.3 for the private stage). But the parallels are not reassuring, and nothing in the following narrative bears this out.

3. The *scaena* is metaphorical, meaning he made a show of his immorality instead of keeping it private (cf. 1.16.3 for his behaviour in public baths): cf. *Ep.* 94.71 *ambitio et luxuria et inpotentia scaenam desiderant: sanabis ista si absconderis.* Cf. L&S s.v. *scaena* II.1-2; *OLD* s.v. 4; Otto 311 s. *scaena*; Cic. *ad Brut.* 1.9.2 ... *tibi nunc populo et scaenae, ut dicitur, seruiendum est ...*; *de Or.* 3.162 *quamuis sphaeram in scaenam, ut dicitur, attulerit Ennius*; 177 *neque ex alio genere ad usum cotidianum, alio ad scaenam pompamque sumuntur*; *Planc.* 29; Hor. *S.* 2.1.71 *ubi se a uulgo et scaena in secreta remorant*; Ov. *Tr.* 1.9.48 *scaena manet dotes grandis, amice, tuas.*

Shackleton Bailey objects to the *usque*, for where else except on stage could the thing be 'produced'? With interpretation 2 this is difficult, but with 3 the *usque* could mean that he brought his behaviour right out into the open rather than keeping it private and little known. So I do not think that Shackleton Bailey's *canitiem*, or Watt's palaeographically preferable *senium*, is necessary.

A further question is whether S. is playing on the supposed derivation of *obscenus* from *scaena.* (Corcoran 82 n. 1 says there may be a play on words, but does not mention the etymology.) Cf. Var. *L.* 7.96 *obscaenum dictum ab scaena ... turpe ideo obscaenum, quod nisi in scaena palam dici non debet.* This particular etymology, however, does not fit this context, and so could not readily have been in S.'s mind here.

As Axelson[1] remarks, there is nothing wrong with *perductae in scaenam* in sense 3 (though *producere in scaenam* in sense 1 is found, cf. Suet. *Ner.* 4).

1.16.5 quem non putes in ipso habitu pingi uoluisse?

quem Zθπ: quam R: quid δ   ipso Ψ: illo ζ   pingi Ψ: et pingi ζ   noluisse V

ζ's *illo* and *et* are preferable (if we read *ipso*, then *habitus* has to mean virtually 'sexual position', but *TLL* 6.3.2483.35-41, on *habitus* of a position of the body, gives no parallel for that). A connective relative *quem* is unlikely at the start of a question, where it could be mistaken for an interrogative (so we may rule out F. Muller's *quem nonne* ..., a combination for which I have found no sure parallel; but Professor Watt points to Val. Max. 9.13.1 *quem non* (*non*<*ne*> Perizonius: om. Γ Halm) *aliquis merito dixerit* ....?; Stangl, *BPhW* 32 (1912), 1557, defends *quem non*). Oltramare turns it into a relative clause: ..., *quem non putes in ipso habitu pingi noluisse*. But a more satisfactory approach is offered by Brakman[1]: *quid? eum non* ..., except that the *eum* is placed too emphatically. Preferable would be *quid? tu eum non* ...; cf. *Ep.* 66.13 *quid? tu non putas* ..., *Dial.* 2.4.2 *quid? tu putas* ..., *Ben.* 6.32.3 *quid? putem* ..., *Ep.* 67.11 *quid? tu existimas* ... (and *quid? tu* ... *iudicas* ... at *Ben.* 7.16.3, *Ep.* 95.25, *Nat.* 2.32.7). (Other proposals are unattractive: Walter[2]'s *quem* <*non est quod*> *non putes* ... is too cumbersome; Alexander's *quemnam putes* ... ('who in the world ...') is entirely the wrong sense. Vottero[3]'s *quem num putas* ... *noluisse* loses the potential. Gercke's *eumque non* (in app.) is tolerable, Leo's *quem cur non* is bland and further from the manuscripts. Kroll[1] suggested *quidni putes*, but surely then (as with Leo) *eum* or similar is needed?)

1.16.7 etiam ea quae a conspectu corporis nostri positio submouit ...

a *om.* ζ

In S.'s prose the ablative with *submoueo* is always accompanied by *a(b)* (e.g. *Nat.* 1.2.11, 1.3.9, 2.10.4), so ζ should not be followed here.

1.16.7 nihilominus illa quoque superuacua mihi parte alicuius contumelia marem exerceo.

In place of *alicuius* Leo suggested *alienaque a*, and Gercke in the apparatus apparently *obtusus ad alienam* (presumably with *contumeliam*?). Housman[2] 405-6 = Housman[4] 1178 explained that *marem ex-*

erceo means 'I use my virility', and *alicuius contumelia* refers to the fact that the other person is an *irrumator*. Vottero[3] conjectures *alicuius contumelia mar<is> m<e> exerceo*, objecting that since Hostius is effeminate he cannot be said to exercise his virility. But S. is here putting direct speech into Hostius's mouth, and the phrase is appropriate from his lips.

1.17.3 praeterea duorum siderum occursum quo interpellari dies solet non uideremus, nec scire possemus quid esset quod solem nulla obuersante nube subduceret, nisi liberius humi solis lunaeque imagines uideremus.

interpellari ζ, *ut coni. Schottus*: interpolari Ψ   quod solum - subduceret *om.* Ψ   quod Z: quam R   nulla Z: nullo R

ζ's *interpellari* (conjecture by Schottus) and *quod solem nulla obuersante nube subduceret* is clearly right (see Parroni[2] 167-8; Watt).

liberius humi is difficult, but I suspect we should accept it. We may expect mention of a reflection in water or other liquid, not just 'on the ground'; but perhaps *humi* is in implicit contrast to 'in the sky'. It is doubtful whether *liberius* can mean 'plus aisément' (Oltramare); rather 'più liberamente' (Vottero) (cf. *OLD* s. *libere* 2, 'without restriction, at will, freely').

Watt's *tolerabilius* for *liberius* has its attractions, but is not essential. Older conjectures are not attractive: *nisi laeuore huiusmodi* Pincianus; Gercke in the apparatus suggests *liberi* for *liberius*; F. Muller suggests *nisi liberius lumine (h)umido* (i.e. *lumine aliquo umore hebetato uel lenito*).

1.17.6 et mox huic proprie ministerio praeparatus est orbis, nondum argenti nitor sed fragilis uilisque materia.

praeparatus Ψ: formatus ζ   argentei E   nitor sed RΨ: nitore sed Z: nitoris E

Perhaps *formatus* is a gloss to show that *praeparatus* here means not 'got ready' but 'made'.

The phrase *nondum argenti nitor sed fragilis uilisque materia* presents problems of grammar and sense. As it stands the phrase is in bald apposition to *orbis* (translators paper over the harshness of this). E's *ar-*

gentei nitoris sed for *argenti nitor sed* would give a smoother construction, if one also wrote *materiae*, following Watt. (Z's *nitore* is probably accident, since R has *nitor.*) But the sense is also problematic. The *fragilis uilisque materia* contrasted with silver is presumably pottery (so Gercke in his apparatus, and Alexander); but no parallel is given for the use of glazed pottery in mirrors, not even in ancient speculations about primitive times, and it is scarcely to be understood from the Latin. Furthermore, pottery mirrors, even if they existed, do not fit the progression of thought in the section. The development of man's interest in his own reflection is traced in four distinct stages in §§ 5-6: (1) people accidentally caught sight of their reflections in pools, etc.; (2) they deliberately looked at such reflections of themselves; (3) after metals were discovered, they caught sight of their reflections in cups or other bronze objects not intended as mirrors; (4) they deliberately made mirrors. The mirrors in (4) ought to be made of bronze or other metals, for there is no need for people at this stage of development to revert to pottery, even supposing it was suitable for mirrors.

In fact the phrase perfectly well describes a contrast between silver objects and pottery ones, but this is the wrong context for that contrast. If *fragilis uilisque* is sound, then either the phrase is Seneca's, but has been transposed from elsewhere, or it is interpolated (but the clausula counts against interpolation). A similar contrast is in question in §§ 8-9, talking about the simplicity of the early Romans, but there is nowhere in that passage which can accommodate, never mind which demands, our phrase; nor will it fit in §§ 5-6.

Another approach would be to find a replacement for *fragilis*, so that the reference could be to bronze as opposed to silver. Watt suggests *parabilis*; but this still leaves the doubt whether *uilis* could be used of bronze in this context.

1.17.10 processit enim paullatim in deterius opibus ipsis inuitata luxuria.

Geist[1] 34 finds *inuitata* too strong, and suggests *inritata*, which is accepted by Vottero. But cf. e.g. *Ep.* 76.18 *(uir bonus) ab honesto nulla re deterrebitur, ad turpia nulla inuitabitur*; 123.13 *quod adsequi poterimus si scierimus duo esse genera rerum quae nos aut inuitent aut fugent. inuitant [ut] diuitiae, uoluptates, etc.*; *OLD* s. *inuito* 6.

2

Notes on *Naturales Quaestiones* Book 2

2.9.2 hoc (spiritu) attollitur (aqua) inserto et cogente; contra naturam suam multa conatur et ascendit, nata defluere.

Watt is uneasy about *multa conatur*, and proposes *sursum conatur*; but one may compare 1.2.9 *itaque ex hac inconstantia caeli tam multa temptantis et undique laborantis* ...; 6.21.1 *nobis quoque placet hunc spiritum esse qui possit tanta conari* ...

2.21.2 quid, quod omnibus fulguratis odor sulphuris est?

fulguritis *Fortunatus*

Elsewhere in classical authors transitive *fulgurare* is only doubtfully attested: Plin. *Nat.* 15.57 *quando fulgurata* [M: *fulgura* Ead, fort. recte: *fulgora* DFe] *piari non queunt facile*; Serv. Auct. *A.* 12.169 *fulgurata (uestis)* [*fulgorata* F: *fulgurita* T]. Transitive *fulgurare* is found in the Vulgate and Christian Latin (*TLL* 6.1.1521.73-6), and transitive *fulgorare* is found at *CGL.* 5.458.4 *Fulgoratus fulmine percussus*; 5.569.27 *Fulgoratus fulgore percussus*. On the other hand transitive *fulgurio* is a well-attested religious term (see *TLL* and *OLD* s.v., including S. *Dial.* 5.23.6). So Fortunatus may be right.

2.26.4 spumabat interdiu mare et fumus ex alto ferebatur; †non† demum prodebat ignem ...

non ζ: nam Ψ: nox *Madvig*[2] non demum] nocte unda *Hine*[6]

I was wrong to say (Hine[6]) that temporal *demum* cannot be used with the imperfect: cf.—with various uses of the imperfect—Curt. 3.11.6 *tum demum ergo promouebant gradum, cum hostem prostrauerant*; 7.3.14; Plin. *Nat.* 36.45 *columnis demum utebantur in templis*. Here S. could perfectly well be using the imperfect to describe a sequence of events that occurred on several successive days and nights, so Madvig's conjecture is the simplest.

2.34.4 non refert an aliud sit per quod quidem quaerimus, quoniam de quo quaeritur idem est.

quidem θPU: *om.* ζL²δW¹: quid W², *ut coni. Castiglioni¹:* uerum *uel* quidem uerum *Watt*

Goodyear[2] asks where *quidem* could have appeared from; a possible answer is by dittography of *quod* abbreviated, with consequent alteration to make some kind of sense. However, in advocating the omission of *quidem*, I ought not to have said (Hine[6] ad loc.) that W² and Castiglioni's *per quod quid[em] quaerimus* is not Latin: for indefinite *quis* in relative clauses see *Ben.* 1.12.3 *quanto acceptius est si id damus quod quis non habet, quam cuius copia abundat?*, 3.19.1 *beneficium enim id est quod quis dedit cum illi liceret et non dare*, 5.11.6, *Ep.* 18.9, and *OLD* s. *quis*² 1b; K-S 1.633-4 is less helpful on relative clauses (and one might also note further extensions of the use of indefinite *quis* in S. *Nat.* 5.18.11 *non immerito quis dixerit*, and 6.23.3 *quotiens quis dixerit*).

One might well wonder whether *per quod quid quaerimus* offends against euphony; but it is followed directly by the hardly less striking *quoniam de quo quaeritur*, and compare e.g. *Nat.* 3.7.3 *quid quod quaedam flumina erumpunt ...*, *Dial.* 10.7.10 *non est itaque quod quemquam ...*, *Ben.* 3.2.1 *quis tam ingratus est quam qui, quod in prima parte animi ...*, 3.9.1 and 3.10.3 *quid, quod quaedam ...*, 5.7.1 *quod qui quaerit ...* So *quid[em]*, accepted by Vottero, is the best reading in this passage, the corruption being very easy.

2.43.2 discant hi, quicumque magnam inter homines potestatem adepti sunt, sine consilio ne fulmen quidem mitti.

potestatem adepti sunt ζ: sapientiam a- s- A: a- s- sapientiam θπ: potentiam a- s- BV

Hine[6] tentatively printed ζ's reading, Vottero reverts to Gercke's *adepti sunt potentiam.* My previous discussion ignored one factor: it could be argued that *potentiam*, with its pejorative tone, is preferable to *potestatem*. However, we should note that the rhythm is better with *potestatem* (*-tātem ădēptī sūnt*); and S. elsewhere uses *potestas* where *potentia* might have been expected, for the sake of rhythm: cf. *Clem.* 1.11.4 *quid enim est cur reges consenuerint liberisque ac nepotibus tradiderint regna, tyrannorum execrabilis ac breuis potestas sit?*, *Dial.* 5.13.7 ... *discere licet, quantum mali habeat ira ubi hominum praepotentium potestate tota utitur,* ... (admittedly, apart from considerations of rhythm, S. may have avoided *praepotentium potentia*); and with our passage compare *Clem.* 1.8.5 *ut fulmina paucorum periculo cadunt, omnium metu, sic animaduersiones magnarum potestatum terrent latius quam nocent, non sine causa* (the plural of *potentia* is not generally used, of course). Conversely *potentia* can be used without invidious connotations, as at *Clem.* 1.20.3 (maybe for rhythmical advantage) ... *qui intellegit magni animi esse iniurias in summa potentia pati nec quicquam esse gloriosius principe inpune laeso*; cf. 1.13.4 (rhythmically neutral) ... *in cuius animo nihil hostile, nihil efferum est, qui potentiam suam placide ac salutariter exercet adprobare imperia sua ciuibus cupiens* .... In any case in the present passage S. did not necessarily want any pejorative overtones. So ζ's reading remains the most plausible.

2.54.1 e terra terrenisque omnibus pars umida efflatur, pars sicca et fumida; haec fulminibus alimentum est, illa imbribus.

Watt supplies genitive *halitus* before the first *pars* to make meaning and construction clear. The sense is arguably clear enough, but if supplement is needed, it is not ideal to introduce the key term in the genitive dependent on *pars*. It is preferable to supply something like ... *terrenisque omnibus <effertur halitus>: pars umida* ... Various other verbs are possible (*fit*, *exspiratur*, *emittitur*, etc.)

3

Notes on *Naturales Quaestiones* Book 3

3.pr.4 festinemus et opus nescio an superabile, magnum certe, sine aetatis excusatione tractemus.

an <in>superabile *Kroll*[2]

Is Kroll's supplement needed? First, can *nescio an* ... mean 'I rather think ... not', 'probably not', in S.? H-Sz 543 cites Sen. *Contr.* 9.5.15 *nescio an quicquam melius*, Val. Max. 5.2.9 *multa insignia Pompeio, sed nescio an hoc quicquam admirabilius contigerit, quod* ..., and later examples. But in S. elsewhere *nescio an* means 'I rather think, probably', cf. *Ep.* 117.20 *tam necessariam curam ... in re nescio an falsa, certe inutili terimus?*, *Dial.* 2.4.3 *immo nescio an magis uires sapientiae ostendat tranquillitas inter lacessentia, sicut maximum argumentum est imperatoris armis uirisque pollentis tuta securitas in hostium terra*, 8.4.2, 12.12.1, *Clem.* 1.6.2, *Ep.* 39.2, 55.1, 88.32, 108.16.

It should be noted that the elder Seneca and Valerius Maximus frequently use *nescio an* to mean 'probably' (e.g. Sen. *Contr.* 1.pr.21 *et nescio an maximum uitium subtilitatis sit nimis se ostendere*, 1.2.6, 1.4.4, al., Val. Max. 1.1.14 *nescio an omnes M. Atilius Regulus praecesserit*, 2.1.6, al.). They use *nescio an* to mean 'probably not' only where it is followed by *quicquam* (as in the two examples quoted above) or a similar word (Sen. *Contr.* 7.pr.2 *splendor orationis quantus nescio an in ullo alio fuerit*). Cf. below on 6.1.11.

3.pr.11 quid enim [dignum] habet quod concupiscas?

habet *Skutsch*: dignum habet Ψ: habes Zγ

Recent editions have followed Skutsch, but should have followed Z, for *habet* (sc. *fortuna*) is less general than *habes*. Perhaps they did not recognise the idiom: cf. *Ep.* 30.4 *nil habet quod speret quem senectus ducit ad mortem*; Cic. *Sen.* 68 *senex ne quod speret quidem habet*, *Att.*

9.12.4 *nulla re iam possum iuuari, qui ne quid optem quidem iam habeo nisi ut aliqua inimici misericordia liberemur*; Sen. *Contr.* 7.pr.8 *quid habeo quare ...*; *TLL* 6.3.2437.13-28.

3.pr.11 ... quotiens ad humana recideris, non aliter caligabis quam ...

Rhythm will be improved by spelling *reccideris* (cf. *Dial.* 4.36.4, 5.39.3, 10.12.3, *HF* 269).

3.1.1 Quaeramus ergo de terrestribus aquis, et inuestigemus qua ratione fiant—siue, ut ait Ouidius,

fons erat illimis nitidis argenteus undis,

siue, ut ait Vergilius,

unde per ora nouem uasto cum murmure montis
it mare proruptum et pelago premit arua sonanti,

siue, ut apud te, Iunior carissime, inuenio,

Eleus Siculis de fontibus exilit amnis,—

si qua ratio aquas subministraret, quomodo tot flumina ingentia per diem noctemque decurrant, quare alia hibernis aquis intumescant, alia in defectu ceterorum amnium crescant.

ut ait Ouidius - undis, siue *om.* Z   proruptum Z, *Vergili codd.* *$F^1M^1RV^1$ω, 'melius' iudicat Seruius*: preruptum Ψ, *Vergili codd. F^2M^2γV^2adr, Tib. Cl. Donatus*   eleus B, *ut coni. Pincianus*: Elius *Gercke*: elisus ZH: elisius AVFπ   si qua Zθπ: que δ: quomodo ς: siue qua *uel* seu qua *Gercke ap. Gercke-Norden 54*: *fort.* siue qua <alia> subministraret ZθρU: subministrat W^1Cam.E: subministret δ

Z has *proruptum* and Ψ *praeruptum*, mirroring the variants of the ancient and mediaeval Vergilian tradition, and possibly contaminated from Vergil manuscripts—but not necessarily, for the variants could easily have occurred independently in the Senecan tradition.

ZH's *elisus* is at first sight possible (for *elido* of water cf. *Nat.* 1.3.2, 6.31.3, 3.15.7), for it is appropriate to the fast-flowing Arethusa (and the rest of the line can scarcely refer to anything else); however, it overlaps in sense with *exilit*, and *Eleus* is more pointed, being a proper name, and alluding to the legend of Arethusa's origin in the Alpheus. Furthermore, what is the function of these three quotations? Do they describe three different categories of terrestrial water? If so, the first would be a small, quiet spring, the second a huge river, the third—reading *Eleus*—a river that flows underground and re-emerges. (Some ancient sources said that the Timavus flowed underground, but not Vergil; see Str. 5.1.8, C214-5, and Kidd on Posidonius F225 Edelstein-Kidd; *RE* 6A s. Timavus.) With *elisus*, the point would have to be its forceful emergence, but then *de fontibus* seems poor; *de faucibus* or the like would better suit the force of its appearance. (The form *Eleus*, as found in the plays at *Ag.* 918, *HF* 840 (the text is uncertain at *Oed.* 285), is preferable to Gercke's *Elius*; for *Elius*, the name of a *leno* at *Dial.* 1.5.2, is not relevant, and not secure, for it could be *Aelius* (*ęlius* A)).

The paradosis is *si qua ratio aquas subministraret.* Not only does the verb ending need changing (*subministrat* is usually accepted), but even then the clause is unsatisfactory. As Gercke explains, it must mean 'nisi ratio prorsus abest', but it is unlikely that S. would ever say such a sceptical thing; and even if he did, would he do so here, in the middle of a series of indirect questions depending on *inuestigemus*? There are various possibilities:

(1) one might suggest *[si] quae ratio aquas subministret*, to make another in the series of indirect questions; but this question would not be sufficiently distinct from *qua ratione fiant* at the start of the sentence;

(2) following Gercke's suggestions of *siue qua* (but not *seu qua*—*seu* is only very shakily attested in S.'s prose, see below on 3.25.11), one might suggest *si<ue> qua <alia>*, so that *ratio* here picks up *qua ratione* early in the sentence;

(3) the whole clause might be a gloss, perhaps generated from a marginal paragraph heading *quae ratio aquas subministrat*—though is *subministrat* a glossator's word?

Perhaps (2) is best: *quomodo* then introduces a different question, not about the sources of individual rivers, but about the enormous amount of water needed to keep all the world's rivers supplied.

3.2.1 Aut stant omnes aquae aut eunt aut colliguntur aut uarias habent uenas. aliae dulces sunt, aliae uarie asperae ...

Codoñer Merino is right that *Aut stant - habent uenas* does not belong here, and her transposition to the start of ch. 3 (before *Vt stet aqua* ...) is plausible: skipping from *Aut stant* to *Vt stet* could account for the initial omission. (Alternatively one might wonder if it is a marginal section heading that has crept into the text at the wrong place: but it is very rhythmical, *Aūt stānt ōmnēs ăquaē/ aut eunt/ aut cōllĭgūntūr/ aut uarĭās hăbēnt uēnās*, so less likely to be an interpolation.)

3.3 ut stet aqua aut fluat, loci positio efficit: in deuexo fluit, in plano et deuexo continetur et stagnat.

et deuexo θπ: *om.* δ: aut supino Z: et conuexo *Leo, Pusch ap. Gercke*: et depresso *Gercke in app.*: et deiecto *Kroll*[1]: et supino *Oltramare*: et defosso *Vottero*[3], *Codoñer Merino*

Ψ has repeated *deuexo* from the preceding phrase, ousting the true reading. Z's *aut supino* is possible, for *supinus* can mean low-lying, and it is especially used of gently sloping land (cf. *OLD* s.v. 3, Oltramare ad loc.); a good parallel for our passage would be Plin. *Pan.* 30.4 *neque enim solum uagus ille* (sc. Nile) *<cum> expandatur amnis intra usurpata semper collium substiterat atque haeserat, sed supino etiam ac detinenti solo non placido se mollique lapsu refugum abstulerat ...* Z's *supino* would not be an obvious interpolation, so it should be accepted.

3.5 colaturque in transitu mare, quod per multiplices terrarum anfractus uerberatum amaritudinem ponit et prauitatem in tanta soli uarietate saporem exuit et in sinceram aquam transit.

saporem ZθρU: uaporem W[1]: saporis (*ante* in tanta) δ: saporum *Skutsch*: *del. Fickert* in (*ante* sinceram) *om.* Z

In the latter part of this sentence the punctuation is problematic, and so has been omitted above. Also, *uerberatum* is suspect, because normally water (or wind) beats something else rather than being beaten itself (cf. of water 4a.2.5, 6.7.4, *Dial.* 7.27.3, *Tro.* 1122, *Pha.* 6, *Oed.* 10; of wind, *Nat.* 4a.2.23, 4b.8, *Ep.* 83.7), and the same goes for *euerberatum*, the conjecture of Castiglioni[1], printed by Vottero (cf. *Nat.* 6.27.3; *OLD* s.vv. *euerbero, uerbero* 3b). 5.12.2 is maybe more help: *his interuallis tenuis includitur spiritus qui maius desiderat spatium cum euerberatus cursu parum libero incaluit.* But even granted that the sea can be buffeted as it passes, how is that meant to remove the saltiness from the water? Hence the initial attractiveness of C. F. W. Müller's *percribratum*; or perhaps *cribratum* is as easy palaeographically. But the difficulty is that *cribro* is properly used only of sifting dry substances, and *colo* (used just above by S., so not likely to be repeated here) is the word for straining liquids. However, for the two ideas combined cf. Colum. 7.8.7 *thymum contritum cribroque colatum*; perhaps S., having just used *colatur*, did use *(per)cribro* in the same sense.

With the paradosis *saporem*, the end of the sentence can be made intelligible with punctuation after *prauitatem* (so Oltramare edn. 1.162). But Fickert's deletion of *saporem* (as a gloss on *prauitatem*) is attractive. However, in any case, what has variety of soils got to do with the sea water losing its saltiness? If anything, variety of soils should give variety or complexity of flavour, not remove an objectionable flavour. Could *in tanta soli uarietate* be a gloss on *per multiplices terrarum anfractus*? But perhaps the text is sound and the point is that if the sea water passes through a wide variety of different soils, they will sift out different impurities, and it is bound to lose its saltiness somewhere or other.

3.7.2 quomodo ergo imber suggerere potest amnibus uires, qui summam humum tinguit? pars maior eius per fluminum alueos in mare aufertur; exiguum est quod sorbeat terra, nec id seruat: aut enim arida est et absumit in se quicquid infusum est, aut satiata si quid supra desiderium cecidit excludit, et ideo primis imbribus non augentur amnes, quia totos in se terra sitiens trahit.

J. Müller[1] 19-20 was right to say that *pars maior ... sitiens trahit* belongs to an interlocutor: for the previous part of the chapter is challenging the view that rain water is the source of rivers, these lines are supporting it, and §3ff. again challenges it. Müller inserts *inquit* after *pars*, but it is palaeographically easier to insert something like <*'quoniam' inquit*> (lost after *tinguit*) before *pars* (cf. 1.7.1 *'quoniam' inquit, 'uitri ...'*). Professor Watt suggests, more simply, *'<at> pars ...* (again readily lost after *tinguit*; for this use of *at* in S. cf. e.g. 3.10.2 *cur non aqua fiat e terra? 'at magna flumina sunt.'*).

3.8 Quidam existimant, quemadmodum in exteriore parte terrarum uastae paludes iacent magnique et nauigabiles lacus, quemadmodum ingenti spatio maria porrecta sunt infusa uallibus, sic interiora terrarum abundare aquis dulcibus, nec minus illas late stagnare quam apud nos oceanum et sinus eius, immo eo latius quia plus terra in altum patet.

late Z, *post* stagnare *add.* L²: *om.* Ψ   quia ZδθPυ: quod R: quo T

quia is the paradosis, and found in the earliest editions, but *quo* is found in all editions since Fortunatus. However, for *eo* plus comparative plus *quia* cf. *TLL* 7.2.477.68-70; e.g. Plaut. *Most.* 763-4 *eo maiore ... opere ... quia ... audiuit*, Cic. *Phil.* 11.24 *quod quidem eo uehementius tibi laborandum est quia in eandem suspicionem cadere collega ... non potest.* In fact in our passage *quo* gives the wrong sense, meaning 'indeed the water spreads out more widely the deeper the earth extends', as though the depth of a spherical earth admitted of variations at different points. S.'s thought is compressed and hard to follow, but possibly he means that the depth of the seas and the ocean is only a tiny fraction of the earth's diameter, so the great depth of the earth allows there to be bodies of water underground much larger than any we see on the surface. (The thought would be easier, perhaps, if we had ... *nec minus illas alte stagnare ... immo eo altius quia plus terra in altum patet* (for confusion between *altus* and *latus* cf. 4a.2.8 *latitudinem* Fortunatus (*alt-* Ω), 2.26.5 and Axelson[1] 34), but *uastae* and *porrecta* support *lat-*.)

3.9.3 hoc quoque quicquid efflauit, quia non libero aëre concipitur, crassescit protinus et in umorem conuertitur.

concipitur Ψ: excipitur Z

excipitur gives better rhythm, and for the sense cf. *TLL* 5.2.1252.4-11.

3.10.1 quae si in alia mutabilis, est etiam in aquam, immo maxime in hanc.

The ellipse of *est* in the conditional clause is awkward, for in an unpunctuated text it would be initially unclear that *est* goes not with *mutabilis* but with what follows. Maybe another *est* should be supplied after *si* (*est* omitted after *esi*; it should not be supplied after *mutabilis* and before *est*, for that would spoil the clausula *ălĭă mūtābĭlĭs*). Charney 47 gives instances of ellipse of *esse* in relative clauses (e.g. *Ep.* 43.2, cf. 85.2) and *quia*-clauses in the Letters, but gives none of ellipse in a conditional clause (cf. generally K-S 1.11-2, H-Sz 421, but their examples are not as awkward as the present one).

3.10.3 omnium elementorum alterni recursus sunt.

Maybe *alterni <cursus et> recursus*? Cf. *Dial.* 12.20.2 *cursusque eius alternos et recursus*, *Ep.* 88.26 *cursus et recursus*; cf. Mela 3.55 *ob alternos accessus recursusque pelagi.* The paradosis could be defended from Ou. *Ibis* 419 *per alternos unda labente recursus.*

3.10.3 et natura partes suas uelut in ponderibus constitutas examinat

Here *ponderibus* needs to mean 'balance, scales', and that is the meaning given at *OLD* s. *pondus* 3b. But of the other passages there cited, at *CIL* 10.1453 *pondera* seems to mean weights, not scales (cf. line 6 *ut uitieis ponderum occurrerint*), and in *CIL* 14.375.29 it is not clear that it means balance as opposed to weights (*[id]em pondera ad macellum cum M. Turranio sua pecunia fecit*). L&S s. *pondus* IBb give *pondera* = 'balance, equipoise, equilibrium' (abstract), also citing Vulg. Isa. 40.12 *quis ... librauit in pondere montes et colles in statera?* Du Cange s. *pondus* gives 'statera seu bilanx publica'. So the required meaning of *pondera* is not certainly attested before the Vulgate, and there is room for suspicion that *ponderibus* is here a gloss on another word, maybe *lancibus*. S. uses *lanx* (*Ep.* 119.5), but not *trutina*, *libra* (not in the sense 'balance'), or *statera*—which only means he does not often talk of balances. The hypothesis of a gloss is supported by *CGL*: 4.532.51 *lance pondere*, 4.254.4 *lances pondera, mensurae* (cf. the corrupt 5.505.24 *lances discum uel mensura* [= *missoria*, index vol. s. *lanx*] *siue co[n]cinae pondera* [*pondera* add. manus secunda]). *stater* and *trutina* are not glossed with *pondus* or *lanx* in *CGL*.

3.12.3[4] satis et multum illi uirium dedi

Muretus deleted *et*, Garrod[2] proposed *nimium* for *multum*, and modern editors print the transmitted reading. I can give no parallel for it, and *satis multus* is commonplace: S. has *satis multum temporis* at *Ep.* 19.1, cf. 81.3 ... *satis multa ... locuti sumus.* But S. elsewhere plays on the distinction and progression between the words, and that seems enough to justify the paradosis here: cf. *Ep.* 119.6-7 *utrum mauis habere multum an satis?*

etc., *Dial.* 6.21.2 *uno modo multum est quod uiuimus, si satis est*; cf. also *Ep.* 90.16 *satis abundeque* (and cf. *satis superque*).

3.14.2 hanc ueterem et rudem sententiam explode: nec est quod credas in hunc orbem aquam subire per rimas et facere sentinam.

There is something slightly illogical about *nec est*; for *hanc sententiam* covers the whole of Thales' theory, and *in hunc orbem ... sentinam* is one detail of it; i.e. in effect the sentence says 'don't believe any of his theory, and don't believe this bit either'. *non est* would be easier, with the second sentence in explanatory apposition to the first; or maybe something is missing, on the lines of e.g. *nec <uerisimile est terram in aquis natare, nec> est quod ...* A reference either to the earth floating or to earthquakes may have dropped out.

3.15.7 saepe enim inania adposita plenis umorem in se auocauerunt.

auocauerunt (eu- F[1]) F[c]HPU: aduocauerunt ZR: attrahunt δ: auocauerit W

For *auoco* cf. Vitr. 8.1.7 *ex his (aquis) quod est leuissimum tenuissimumque et subtili salubritate aër auocans dissipat in impetum caeli*, Plin. *Nat.* 17.118. For *aduoco* of things cf. *TLL* 1.894.70-895.30, in the sense *auxilio arcessere, adhibere*, e.g. S. *Nat.* 1.16.4 *oculos suos ad illa aduocauit*, *Ep.* 78.5 ... *uinum quando uirium causa aduoces.* But this sense is not so appropriate here.

3.15.8 haec tenuis unda uix fonti sufficit: et ex magnis causis magnisque conceptibus excidunt amnes, nonnumquam leuiter emissi, si aqua tantum pondere se suo detulit, nonnumquam uehementer et cum sono, si illam spiritus intermixtus eiecit.

et (*ante* ex) Zδθυ: uel ρ: at *Pincianus*: sed *Gercke in app.*: *del. Rossbach*[5] causis Ω: cauernis *Diels ap. TLL 3.630.8-11*: caueis *Diels ap. Gercke (sed coni. Gercke, teste TLL)* excidunt Zθπ: excedunt δ: exeunt *Haase III p. xxv*: exiliunt *C. F. W. Müller* leuiter (*plura om.* δ) Ω: leniter *Shackleton Bailey*

The syllepsis *ex magnis causis magnisque conceptibus* is harsh, but could be understood as making a distinction between the efficient and material causes; for the first part cf. *Ep.* 77.4 *ex maximis causis*; and *Nat.* 5.8.3 *prout (aura) ualentioribus minoribusue collecta causis est*; 5.12.3 *si (uenti) alios quoque flatus ex eadem causa ruentes in se abstulerunt.* But general reference to efficient causes is hardly apt in this context, and emendation is needed. Modern editions print *caueis* (conjectured by Diels according to Gercke, but by Gercke here and elsewhere according to *TLL*, see above), but the conjecture is doubtful: the meaning 'hollow, cavity' is alleged by *OLD* s.v. 5 to be found at Plin. *Nat.* 11.3: *(insectis) qua subtilitate (natura) pinnas adnexuit, praelongauit pedum crura, disposuit ieiunam caueam* [EFR: *cauam* D] *uti aluum, auidam sanguinis et potissimum humani sitim accendit!*; but *cauea* can there mean 'enclosure, container' (cf. *OLD* s.v. 1, 3). *TLL* s. *cauea* 3.630.8-15 quotes this passage and the conjecture *caueis*; otherwise the earliest example of this sense is Lact. *mort. pers.* 49.5 *exilierunt oculi eius de caueis.* So it seems that if conjecture is needed *cauernis* is safer, and palaeographically just as simple (*cau(er)nis* → *causis*).

et is not tolerable: Alexander says *et* means 'and yet', but a stronger adversative is needed. *at* is acceptable.

Löfstedt[2] 74 defends *leuiter* on grounds of rhythm. For *leuis* of the flow of water see *TLL* 7.2.1205.4-12, e.g. S. *Ag.* 432-3 *unda uix actu leui tranquilla ... tremit*, *Pha.* 13-4 *per graciles leuis* [*breuis* Axelson] *Ilisos labitur agros*; but *leuis* in those passages is associated with speed, which is not in question here. Here, however, *leuiter* may indicate that the water is emitted without much force, as the following clause explains. Still, *leniter* and *leuiter* are so often confused in manuscripts that Shackleton Bailey's conjecture deserves consideration if the objection to the rhythm can be met. Parallels can be found for the clausula –⏑⏑––x: e.g. 1.2.7 *-ens aqua disturbat*, 1.3.9 *-i speciem reddit*, though in the present passage a word begins at the first longum. With *leniter* one might change to *missi* to improve the clausula. Cf. *TLL* 8.1162.38-41 and 5.2.500.18-29 on confusion between *emitto* and *mitto* elsewhere. For *mitto* of water cf. *TLL* 8.1176.43-9, e.g. Mela 1.50 *(Nilus) ex desertis Africae missus* (perhaps to avoid elision of *-ae* before *emissus*?), besides poetic examples. On compound replacing simple verbs in the manuscripts, see above on 1.2.9.

3.16.1 ... cum possim eandem causam omnibus reddere.

omnibus δHπ: omnis F: omnium Z: de omnibus Cam.DET

With *causam reddere* (cf. *TLL* 3.668.25-28) the genitive of the explanandum is normal (cf. Var. ap. Gell. 1.18.2 *aliquot uerborum ... reddidit causas falsas*; Cic. *Off.* 1.101; Lucr. 4.503; Ov. *F.* 1.278; Liv. 2.3.5; S. *Nat.* 2.14.1 *cum causas stellarum transuolantium redditis*; Tac. *Dial.* 15; *A.* 6.15 is ambiguous), but for the dative cf. Lucr. 5.620 *non, inquam, simplex his rebus reddita causast.* So perhaps here the less usual *omnibus* is correct, and has been normalised in Z. (*Nat.* 2.46 *uim et causam omnibus dedit* is not quite the same.)

3.16.2 sic aquae interualla habent quibus se retrahant et quibus reddant.

reddant Ω: redeant Cam.DEX: se reddant T: reddantur *Castiglioni*[1]

As Alexander notes, *reddantur* is unrhythmical. Can *se* be understood from the previous clause? Wistrand 74 compares: Cic. *Fam.* 6.6.4 *sed tamen plurimi sunt testes me et initio ne coniungeret se cum Caesare monuisse Pompeium et postea ne seiungeret* (*ne* <*se*> Baiter, *ne se* <*di*>*iungeret* Kleyn); Liv. 23.20.8 *pars (auctores essent) ... adiungendi se ceteris Bruttiis ac per eos dedendi Hannibali* [*dedendi* P: *dedendos* P^2 uel $P^3C^1R^2$ etc.: *dedendis* C: *dedendi se* malit Conway]; Verg. *A.* 1.439-40 *infert se saeptus nebula (mirabile dictu) / per medios, miscetque uiris neque cernitur ulli*; 587 *scindit se nubes et in aethera purgat apertum*; add [Ou.] *Am.* 3.5.29 *illuc se rapuit gregibusque inmiscuit illis.*

3.16.4 crede infra quidquid uides supra: sunt et illic specus uasti, †sunt ingentesque† recessus ac spatia suspensis hinc et inde montibus laxa.

sunt[2] *del.* J^2: *ante* uasti *transp. Castiglioni*[5] *197* ingentesque Zθ: ingensque π: ingentes δ inde Ψ: illinc Z

Editors since Gercke have followed J^2 (*sti* repeated might give rise to *sunt*): but the anaphora of *sunt* with δ's *ingentes[que]* is also possible (rhythm benefits, though at this point the dactylic colon is acceptable); or there could be a lacuna, such as *sunt* <*inmanes cauernae*> *ingentesque ...* Castiglioni's transposition is less attractive, since *specus* is then on its own without an adjective.

Probably *illinc* is normalisation. *hinc ... inde* is rare but attested: *TLL* 6.3.2804.76-2805.26, in Ovid and later poets, in Quint. *Inst.* 12.1.40, Tac. *H.* 4.19, and later prose.

3.17.1 multa hoc loco tibi in mentem ueniunt quae urbane in re incredibili fabulam dicas non cum retibus aliquem nec cum hamis sed cum dolabra ire piscatum.

tibi in mentem Ψ: in mentem tibi Z uenient *Muretus* urbane <ut> Cam.DET in re incredibili θPU: in rem credibilem Z: in re credibili δR: incredibili W^1 fabulam (famu- R) dicas (f- dicam scilicet B) Ψ: fabulae Z retibus Ψ: spatiis Z

Approaches to this crux may be divided into two groups: (a) those that punctuate after *dicas*, taking it with what precedes; (b) those that punctuate before *dicas*, taking it closely with what follows. On either assumption, *non cum retibus - ire piscatum* is an exclamation in the accusative and infinitive (K-S 1.720-1, H-Sz 366; e.g. S. *Dial.* 10.12.8 *esse aliquem qui usque eo deliciis interierit ut an sedeat alteri credat!*).

I take the two approaches in reverse order:

(b) If there is punctation before *dicas*, then the *quae*-clause lacks a verb. Gercke proposed *dicta* (or *quae <poetae iocantur>*) in place of *quae*, and continued *dicta urbane in re incredibili. fabulam dicas, non...* Leo proposed changing *fabulam* to *fabulamur*, followed by a colon (but S. never uses *fabulor*). Neither is convincing, and *dicas* looks very much as if it goes with *quae*.

(a) But if *dicas* is taken with what precedes, the basic problem is that we seem to have two objects for it, *quae* and *fabulam* (and Z's *fabulae* does not help matters: it can hardly be genitive, as Oltramare seems to think, translating—with *ut*—'à propos d'un fait incroyable mis sur la scène'). Numerous conjectures are on offer: *fabulam* has been emended to *fabulosaque* (Fortunatus) or *fabulans* (Schultess[2]) (but, as already remarked, S. does not use this verb elsewhere); others are more drastic (*quae urbane ut in rei credibilis fabulam dicas* (Gronovius[1]), *quae urbane re incredibili ut in fabulam dicas* (Michaelis[2]), *quae urbane, ut incredibilem fabulam, dicas* (Codoñer Merino)), none of which is at all convincing.

Others instead make *fabulam* (uel sim.) part of the direct speech: Warmington proposed a transposition: ... *quae urbane, ut in re incredibili dicas: 'Fabulae! non cum ...*; in similar vein Alexander had proposed ... *ueniunt quare urbane in re incredibili 'fabulam!' dicas*, and Vottero[3] suggested ... *ueniunt quibus urbane in re incredibili: 'Fabulae!' dicas. 'non*

...' But the exclamatory *fabulae!* seems not to be found outside comedy, so it is doubtful whether S. would have used such a colloquial expression (*TLL* 6.1.26.52-4 cites Ter. *Andr.* 224, 553, *Haut.* 336, *Phorm.* 492), and *fabulam!* is doubtful (Plaut. *Pseud.* 754 *em tibi omnem fabulam* is not parallel); and as for Alexander and Vottero, *dico* inserted within a piece of direct speech is found only in poetry (*TLL* 5.1.983.11-33). One might, with Warmington's transposition, propose ... *dicas: 'fabula est* [cf. Z's *fabulae*] *non* ...' (cf. S. *Nat.* 3.26.7 *illic stabulare Solis boues fabula est*, 3.17.3 *hi* [*ubi* Madvig[2]] *sunt qui fabulas putant piscem uiuere posse sub terra*), except that this blunts the edge of the *urbanitas*; in any case, transposition creates a poorer clausula, *incrēdĭbĭlī dīcās*.

This leaves us with Haase's approach, deleting *fabulam*, and inserting *ut* with Cam.DET (Oltramare's statement that *ut* is in Z is incorrect): *quae urbane ut in re incredibili [fabulam] dicas*. But the sense of the postulated text seems perfectly clear, so why should *fabulam* have been inserted? It occurred independently to Professor Watt and myself that maybe *re incredibili* is a gloss on an original *fabula*, intended to specify which of the various nuances of *fabula* was in question here: *quae urbane ut in [re incredibili] fabula dicas*. Compare *CGL* 5.533.68, where *fabule* is glossed *ineptie*, which is almost certainly (as the *CGL* index s. *fabula* says) an elucidation of the Terentian use of *fabulae* mentioned above.

3.18.1 longa somniculosae inertisque luxuriae neglegentia[m] quam sero †expressero† circumscribi se et fraudari tanto bono sensit.

longa Ω: langor *Leo*: languor *Vottero* neglegentia *Gercke, lectionem codici Z perperam attribuens*: neglegentiam Z: *om.* Ψ

Gercke's *neglegentia[m]* (based on Z) should be accepted, since it restores sense without the need to alter *longa*, and it does not look like an obvious interpolation; Oltramare, Corcoran and Codoñer Merino print it. Vottero objects that *luxuriae neglegentia* here would conflict with *sollertiam luxuriae* in s. 3, but there Seneca is talking, whereas here we have the words of the luxury-loving interlocutor whose attitude is different. For *expressero* Erasmus[2]'s *experrecta* (which he took to agree with *luxuria* understood) is excellent, continuing the image of *somniculosae inertisque*; *expressa* is meaningless here. Various conjectures have tried to get anaphora of *sero*, which is attractive, though the conjectures are unacceptable in other respects: Fortunatus, ... *quae tam sero hoc experta, sero enim circumscribi ...*; Ruhkopf, *qua sero expressa, sero circumscribi ...*; Oltramare, *quam sero expressa sero circumscribi ...* (Vottero[3] proposes *quam<quam> sero experrectus*). Neater would be something like ... *neglegentia quam sero exper<recta est, quam> sero circumscribi ...*

3.18.3 iam pro putrido his est piscis occisus. 'hodie eductus est.' 'nescio de re magna tibi credere. ipse oportet †me credas† huc adferatur, coram me animam agat.'

credere. ipse Ψ: ipse credere. Z   me (*ante* credas) Zθπ: me (ut B) <tibi> δ

For *educo* of catching fish see *TLL* 5.2.119.41-45. The problems in the next sentence have been variously diagnosed and treated:

(a) The following conjectures are on offer from those who think only *me credas* needs changing:
Erasmus[1]: *mihi* [or *mi* Vassis] *credam*
Gercke in app.: *ne tibi credas:*
Haase: *ne caedas,*
Kroll[1]: *nec caesus*
Leo: *integer nans*
Gercke in app.: *intrepidus*
Brakman[1]: *mercatus*
Alexander: *ne credas* [sc. *de re magna tibi*].

(b) Others alter *ipse - credas* in various ways:
Schultess[2]: *ipsi oportet* [sc. *mihi credam*]. *ne caedas,*
J. Müller[3]: *ipsi oportet mensae credas.*
Oltramare: *ipsi oportet me credere.*

(c) Madvig[2] rewrites *nescio - adferatur* thus: *nescio; de re magna tibi [credere] ipsi oportet credas; huc adferatur.*

I take various arguments in turn:

(1) Is *nescio de re magna tibi credere* possible Latin, or must we punctuate with Madvig? Madvig says that *nescio credere* cannot mean *non possum (audeo) credere*; but J. Müller[3] offers parallels where *nescio* indicates unwillingness rather than inability. The parallels are not all equally good, but cf. e.g. 5.11.1: *etesiae ... mane nesciunt surgere*; *OLD* s.v. 3a.

Ψ's order *credere. ipsi* is preferable to Z's *ipse credere* (for *ipse* is unnecessary emphasis, as also would be *ipsi* before *credere*).

(2) *oportet ne* + subjunctive is not classical Latin (cf. *TLL* 9.2.742.14). This rules out the conjectures of Haase and Alexander.

(3) The sceptical speaker may say 'I will only believe my own eyes' (so Erasmus' *mihi credam* is possible, and would be improved by Madvig's *ipsi*) but not 'you must believe only yourself', for the other speaker obviously already believes himself; and in this brief dialogue in direct speech the 2nd person singular cannot be a generalised one, 'one should only believe oneself' (which rules out the conjectures of Fickert and Madvig).

(4) We perhaps expect it to be stated that the fish must be still alive when it is brought. *integer nans* (Leo), an adjectival phrase going with *adferatur*, is perhaps on the right lines, though it is an implausible conjecture, and Kroll's *nec caesus* is no better. One might think of various adjectives or phrases (e.g. *incolumis*, *intactus*, *integer*, *adhuc spirans*), but none is compelling, and in any case *ipse* then gives a slightly wrong emphasis, for it is not the fish in person that must be brought, as opposed to a substitute, but the fish alive as opposed to dead.

(5) Perhaps *ipse* refers to the fish, and the thought is that the fish itself must provide the proof: e.g. we should write something like *ipse oportet me doceat*.

ipsi oportet mihi credam (Madvig plus Erasmus) is a strong contender, giving satisfactory sense without excessive emendation; but no solution can be certain, so it is best to obelise *me credas*.

3.19.2 erant enim pinguia et differta ut ex longo otio corpora, ceterum inexercitata et tenebris saginata et lucis expertia, ex qua salubritas ducitur.

tenebris saginata cannot (despite the translators) mean 'fattened in the dark' (that would require *<in> tenebris*, which is maybe correct). It must mean 'fattened by the darkness'. Busche thought that *tenebris* repeated *lucis expertia*, and mention of food was needed, so he suggested *<caeno> tenebris* (but we would need *<caeno in>*) or just *caeno* without *tenebris* (from *Ep.* 95.25 *illa ostrea, inertissimam carnem caeno saginatam*; but oysters are more likely than fish to be thought to eat mud). Watt suggests changing *tenebris* to *taetris* (cf. *Ep.* 5.4). But maybe the paradosis is right and means what it says, that the creatures are fattened by the darkness: farmers did think that animals could be fattened more easily in dark places, cf. Colum. 8.7.1 on fattening hens in a place that is *minimi luminis*, and 8.14.11 on geese, *sint ... calido et tenebricoso loco.* For the active power of subterranean darkness, albeit on inanimate things, cf. S. *Nat.* 6.28.2 *ubi (spiritus) per saecula conditus tenebris ac tristitia loci creuit in uitium.* Busche's argument that *tenebris* repeats *lucis expertia* is invalid, for the latter phrase makes a new point about the unhealthiness of the environment, as the following relative clause shows.

3.19.4 habet ergo non tantum uenas aquarum terra, ex quibus conriuatis flumina effici possint, sed amnes magnitudinis uastae, quorum aliis semper in occulto cursus est, donec aliquo terrae sinu deuorentur, alii sub aliquo lacu emergunt. nam quis ignorat esse quaedam stagna sine fundo?

terrae sinu Z: sinu terrae Ψ   nam *Kroll*[1]: iam Ω

The earth contains within it not only veins of water but whole, enormous rivers (as is proved by the fish found in waters coming from underground). Oltramare and Alexander think that, like the *amnes*, the *terrae sinus* and the *lacus* are also underground. But since the next sentence (whether with Kroll's *nam* or the paradosis *iam*) appeals to common knowledge, this suggests that the previous sentence must also mention visible lakes. There could be at least four categories of underground river:

(1) rivers that are perpetually underground.
(2) rivers that emerge ready made from underground and then stay above ground until they reach the sea.
(3) rivers that emerge ready made from underground but then sink below ground again.
(4) rivers that emerge below lakes.

With the paradosis we find only two kinds of rivers mentioned, category (1) (*quorum aliis semper in occulto cursus est*), with the curious detail that they disappear down holes (*donec aliquo terrae sinu deuorentur*) and category (4) (*alii sub aliquo lacu emergunt*), a quirky selection. Gercke saw rightly that something has fallen out before *donec*, and suggested *<alii aliquatenus aperte* [better *in aperto*?] *fluunt,> donec ...*, i.e. category (3). Maybe more has fallen out, describing category (2). (Gemoll[2], realising that after *aliis in occulto cursus est* we expect the sense 'until they come into the light', suggested *donec aliquo hiatu terrae deuocentur*; but *deuocentur* is the wrong verb (*euocentur* might be better).)

3.19.4 quorsus hoc pertinet? ut appareat hanc aquam magnis amnibus aeternam esse materiam, cuius non tanguntur extrema †sicut flumina fontes†.

hanc aquam] hinc aquae *Gercke in app.*: hanc aquarum *Castiglioni*[1]
sicut Ψ: sic Z   flumina ZBθPυ: fluminum AV: fulmina R

The last clause is difficult. *cuius non tanguntur extrema* might mean (i) 'whose furthest point cannot be reached' (so Oltramare), but it is hard to see the precise meaning or point of this; or (ii) 'whose bottom cannot be reached' (so Vottero, cf. *sine fundo* in the preceding sentence; for *extrema* in this sense see 3.10.1 *utraque grauis (res), utraque densa, utraque in extremum mundi compulsa*, Manil. 3.232 *media extremis pugnant extremaque summis* (and Housman), Plin. *Nat.* 11.179 *summum gulae fauces uocantur, extremum* [*postremum* Ez] *stomachus*); or perhaps we should translate (iii) 'which never comes to an end', i.e. 'never dries up' (so Schönberger and Schönberger; cf. S. *Ep.* 26.1 *inter decrepitos me numera et extrema tangentis*).

But what of the last three words? The paradosis is *flumina fontes*, but most editors read *fluminum fontes*, which certainly improves the rhythm. The preceding context shows that there is a contrast between *flumina* and the sources of huge *amnes*, so we should understand *sicut fluminum fontes* <*tanguntur*>, not <*non tanguntur*>. The general sense ought to be 'which never dries up, as sources of rivers do', or possibly 'whose bottom cannot be reached, as the (bottom of) sources of rivers can be'; i.e. *fluminum fontes* must be a compendious expression for *fluminum fontium extrema*, which is awkward, to say the least. Could *sicut* (originally *scilicet*?) *fluminum fontes* have been a gloss by somebody who misunderstood *extrema* or *materia*, or (as Professor Watt suggests to me) was *sicut fluminibus fontes* meant to be a gloss on *hanc aquam magnis amnibus aeternam esse materiam*?

3.20.3 illinc illud, de quo Ouidius ait

flumen habent Cicones, quod potum saxea reddit

uiscera, quod tactis inducit marmora rebus,

medicatum est et eius naturae habet limum ut corporibus adglutinet et obduret.

medicatum θρ: medicamentum Zδυ corporibus Zθπ: corpora δ

The paradosis is *corporibus*, but most editions accept δ's *corpora*, and translators take *obduret* to be transitive. This last point is dubious: for transitive *obduro*, apart from this passage, *OLD* adduces Cic. *Fin.* 3.37 *quis <se> contra studia naturae tam uehementer obdurauit?*, but Schiche's supplement of *se* is not necessary (probably *obduruit* is correct, see Madvig ad loc.), and *TLL* 9.2.43.20-37 takes us to the Itala, Tertullian, and later writers. The best solution is Cam.DE's *ut corporibus <se> adglutinet et obduret*, as printed by Fickert and Haase, understanding *limus* as subject: 'that it attaches itself to bodies and hardens'.

3.20.4 inde est quod res abiectae in †eundem† lacum lapideae subinde extrahuntur; quod in Italia quibusdam locis euenit.

locum ς

Since no lake has been mentioned, *eundem* cannot be right. Rossbach[6] 1482-3 says that we understand *Puteolanum*, from the previous mention of *Puteolanus puluis*; but was there such a lake? Gertz[2] proposed *eius modi*, which cannot be ruled out, but others have suggested proper names: Gercke in the apparatus *Tyaneum* (cf. Vitr. 8.3.9) 'vel tale aliquid'; Garrod[2] proposed *Euryminaeum* (cf. Plin. *Nat.* 31.29), but *Erymnaeum* would have been better (cf. Plin. *Nat.* 4.32 [*Erymnae*], Str. 9.5.22, C443 [Ἐρυμναί]), Vottero[3] *Velinum* (cf. Plin. *Nat.* 2.226).

But the following sentence, *quod in Italia quibusdam locis euenit*, shows that if a specific lake has just been mentioned, it was not in Italy, or we should require *quibusdam aliis locis*; this rules out *Puteolanum* being understood and *Velinum*. On the other hand if a non-Italian lake was mentioned, we might expect *quod et in Italia ...*, though not necessarily. But a general reference such as Gertz's *eiusmodi* is perhaps right. Obeli are called for. (Alexander proposed *eiusdem*, thinking, implausibly, that *lacus* refers to a basin, tank or tub filled with the water just mentioned; he prefers *adiectae*. Codoñer Merino prints *locum*, but that has no advantages.)

3.20.4 hoc minus tibi uidebitur mirum si notaueris Albulas et fere sulphuratam aquam circa canales suos tubosque durari.

tubosque *Albertus Magnus, Meteor. 2.3.18, ut coni. Muretus (conl. 3.24.3, Ep. 90.25)*: cibosque Zθπ: riuosque δ

tubos is nearer the paradosis *cibos* than δ's *riuos*, and is supported by 3.24.3. It always refers to artificial pipes (*CIL* 10.4842.14 *fistulas canales tubos ponere*, Colum. 1.5.2 *fictilibus tubis*, Plin. *Nat.* 5.128 *tubo coriis facto*, 31.57, Ulp. *dig.* 43.23.1.6, besides S. *Ep.* 90.25), and that is the meaning here, whereas *canales suos* are natural.

3.20.5 nam quemadmodum ebrietas, donec exsiccetur, dementia est, aut nimia grauitate defertur in somnum, sic huius aquae sulphurea uis et habens quoddam acrius ex aëre noxio uirus, mentem aut furore mouet aut sopore opprimit.

aut (*ante* nimia) *Shackleton Bailey*: et Ω   et habens Zθπ: habet δ: habens *Haase*

The paradosis is *et habens*. Perhaps, rather than delete *et*, supply something like ... *uis et <ipsa per se grauis, et> habens* ..., or ... *uis <ipsa per se grauis>, et habens* ...

3.25.7 sunt enim multi pumicosi et leues, ex quibus quae constant insulae in Lydia natant; Theophrastus est auctor.

natant is abrupt, and so is *Theophrastus est auctor.* Elsewhere in *Nat.*, *(x) est auctor* does not stand on its own but is followed or preceded by an accusative and infinitive (3.26.1, 4a.2.13, 16, 5.15.1), and at *Ben.* 6.37.1 we have parenthetic *ita certe Hecaton auctor est.* Garrod[2] proposed *natare*; Watt proposes *natant <ut>*, which is attractive.

3.25.10 itaque etiam si qua in illa saxa sunt, inuenies exesa et fistulosa, qualia sunt quae duratus umor efficit, utique circa medicatorum fontium riuos, quae ubi purgamenta aquarum coaluerunt et spuma solidatur: necessario leue est quod ex uentoso inanique concretum est.

riuos. qu(a)e Ω: riuos, quae *Fortunatus*: riuos, [quae] *Gercke*: riuos. quare *Haase*: <stagna> riuosque *Skutsch*: <ripas> riuosque *Castiglioni*[1] riuos, ubique *Codoñer Merino in app.*

Castiglioni objects to the phrase *fontium riuos*; *TLL* 6.1.1026.47-57 does not have any example. A supplement is the most attractive solution: maybe one could also supply *<ora> riuosque*: cf. 3.30.4 *ora fontium.*

3.25.11 ... siue infusa siue pota sit.

siue[2] Ψ: seu Z, *ut coni. Skutsch*

seu gives a better rhythm; with *siue* it is very trochaic, but possible. But whereas *siue* is found over 150 times in Seneca's prose, *seu* is found only as a variant in a few passages of *Nat.*: at 6.6.1 before *magnum mare* ZθρW have *seu* and δU *siue*, so the former is the paradosis (but before *illud* Ψ has *siue*, Z *sicut*, showing the vulnerability of *siue*); *seu* is also a variant at 2.59.3: *contemne mortem, et omnia quae ad mortem ducunt contempta sunt, siue illa bella sunt, siue naufragia, siue* [ZB: *seu* RAVθπ] *incursus ferarum siue* [Z: *seu* Ψ] *ruinarum subito lapsu procidentium pondera.* So *siue* should be accepted where the two forms are equally well attested.

3.25.11 quod uitium dicunt habere aquam ex rore collectam.

dicunt (habent R) habere (habere *om.* H^1, *post* aquam *habent* H^cW) Ψ: habere aiunt Z

In Seneca *dicunt* is normally used of a specific, clearly identified group of people (e.g. *Nat.* 1.8.2 proponents of a particular theory of the rainbow, 2.34.3 Etruscan fulgurologists, 2.48.2, 7.4.1, *Dial.* 4.13.3, 8.1.4, *Ep.* 8.8, 9.1, 57.7, 65.2, 85.25, 87.39, 94.30, 95.60 (bis), 102.5, 14, 108.38, 117.2; *iudicant* should be read at *Nat.* 2.41.1); in a relative clause it can serve to identify the group (e.g. *Dial.* 5.23.3 *qui ista dicunt*); or it is found with direct speech (*Ben.* 7.4.8, *Ep.* 82.2), where Seneca does not use *aiunt*, although he uses *ait*. But *dicunt* is not used for a general 'everyone says', 'they say', French 'on dit', for which Seneca uses *aiunt* (e.g. *Nat.* 2.30.1, 4b.5.3). So Z's text is to be preferred. (In the passive, where *aio* is of course not available, *dico* is used of general, proverbial statements, e.g. *Ben.* 4.39.2 *illud quod dici solet*, *Ep.* 100.6 *quod dici so-*

let, domus recta est, frg. 121 *ut uulgo dicitur.* Cf. *TLL* 5.1.988.75-989.17.)

3.26.2 (sidera) minus consumunt atque <ex>trahunt.

suppl. Skutsch

Seneca uses *atque* before consonants only for the sake of the rhythm; this is true even before gutturals, where S. avoids *ac*, but uses *atque* only for rhythm (1.3.8, 2.27.2, *Ep.* 41.3, 92.10, 99.20), and otherwise uses *et.* Here, however, *atque* gives a worse rhythm than *ac* would. Skutsch's supplement neatly explains the corruption. However, *extraho* is not the most natural compound in this context: applied to water, it usually describes drawing water from wells, etc. (*TLL* 5.2.2065.24-36), though cf. *extraho* of the sun's action on the Nile in Plin. *Nat.* 5.55-6 *uerum sole per eos dies comminus facto (Nilum) extrahi ardoris ui* ... (56) ... *sed Timaeo si quis extrahi posse credat.* More normal, and palaeographically almost as easy (*ad* lost after *atq(ue)* abbreviated), would be *adtrahunt*, cf. *Nat.* 5.10.4, Lucan 7.5.

3.26.2 ita quod impendio solebat, id incremento accidit.

impendi *cod. Anicianus Dalecampii*: impendii iis *Madvig*[2] incremento] in uicem *Alexander* accedit E

Translators are rather free. The meaning is: 'thus, what used to be added to the expenditure is added to the increase' (understanding *accidere* with *solebat*); for this sense of *accidere* see *TLL* 1.295.68-296.9, e.g. *Nat.* 6.26.1 ... *cum solido liquidum ac molle semper accideret*, where there is no need for δU's *accederet.* Here *impendi* is slightly easier, but not essential.

3.26.3 sub terra uacat locus, omnis autem natura umor ad inferius et ad inane defertur.

The position of ablative *natura*, between *omnis* ... *umor*, is striking, so one might venture to transpose and read *umor natura.*

3.26.8 sed difficilis ratio est quorundam, utique ubi tempus eius rei de qua quaeritur inobseruatum sed incertum est.

[in]obseruatum *Fortunati cod.*: <non> inobseruatum *Oltramare* sed Ω: uel Cam.DEOT: seu *Shackleton Bailey* ut obseruatum sic incertum *Garrod*[2]

Shackleton Bailey is right about the sense required (on confusion of *seu* and *sed* see Housman on Manil. 1.657, note and addenda, and on Lucan. 1.234). But see above on 3.25.11 for doubts whether S. ever used *seu*. Here *siue* is possible (*iue* lost before *inc*), but it gives a less good clausula than *uel* (*(ue)l* was always vulnerable to omission, especially before *i*, and *sed* could be a stopgap).

3.26.8 nam in his quibus cursus est, non possunt uitia consistere, quae secunda uis defert et exportat.

secunda uis Zθ: secunda ρU: secum unda W: sua uis ς

secunda uis could be an extension of *secundo flumine* and similar phrases (e.g. 3.26.5, cf. *OLD* s. v. *secundus* A 2a), but I wonder if S. wrote *secunda uis* <*fluminis*> or the like; though the context probably makes the sense clear enough.

3.26.8 mare uero cadauera instrumentaque et naufragorum reliqua similia ex intimo trahit ...

instrumentaque *Rossbach*[2] *46 n. 30*: stramentaque (stam- P) Ω: ramenta *G. Müller 33-4*: uestimenta *Skutsch* naufragiorum X, *ut coni. Haupt*[2] *449* reliqua similia Ω: reliquiis similia *Erasmus*[2]: reliqua *Gruterus*[2]: reliqua spolia *Castiglioni*[1]

See Axelson[1] on *instrumenta*. The transmitted *similia* cannot be right: *talia* or *alia* is possible, but Castiglioni's conjecture seems best, though to judge from L&S and *OLD* s. v. *spolia* the word never occurs in the context of shipwreck or anything like it. Maybe therefore one should read *naufragiorum reliquias alias* (which after wrong word-division became *naufragorum reliqua si(mi)lia*).

3.27.1 Sed monet me locus, ut quaeram, ...

sed monet θυ: admonet Z: sed mouet ρ

This is the only passage cited at *TLL* 8.1411.82 for *monet locus ut*, but with *admonet* cf. *Ep*. 118.10 *locus ipse me admonet ut* ..., Cic. *Top*. 51 *admonet autem hic locus, ut quaeratur* ... (and for *res admonet ut* generally *TLL* 1.766.68-767.13). Asyndeton is regular at the start of the concluding sections of books of the *Naturales Quaestiones*, cf. 1.16.1 *Hoc loco uolo* ..., 2.59.1 *Intellego quid dudum desideres* ..., 4b.13.1 *'Quid istas' inquis 'ineptias* ..., 6.32.1 *Haec, Lucili uirorum optime, quantum ad ipsas causas; illa nunc* ... (but 5.18.1 is an exception, *Inter cetera itaque prouidentiae opera* ...). So there is a slight preference for *Admonet* here.

3.27.2 ad originem rerum parce utitur uiribus, dispensatque se incrementis fallentibus.

se L^2Hπ: *om.* F: se et Z

If Z's *et* is genuine, one might think of something like ... *dispensatque se et <digerit>*; but a second verb seems otiose, and *et* could come from dittography of the preceding *e*.

3.27.2 quam longo tempore opus est ut conceptus ad puerperium perduret infans, quantis laboribus tener educatur, quam diligenti nutrimento obnoxium nouissime corpus adolescit!

corpus Ψ: his corpus Z

Watt suggests that Z's *his* could be a remnant of *<homi>nis*. Another possibility is *<iuue>nis*, which is the next stage after *infans* and *tener*; though in any case *adolescit* gives the contrast of age.

3.27.5 mox iniuriam et ualidiora sensere: solutis quippe radicibus arbusta procumbunt, et uitis atque omne uirgultum non tenetur solo, quod molle fluidumque est. iam nec gramina aut pabula laeta aquis sustinet.

fluidumque est θρU: fluidumque ZW^1 aut ZθPυ: nec R pabula Ψ: pabulum aut Z <laxatum> *uel* <uictum> *uel* <madefactum> aquis *Shackleton Bailey* aquas *Watt* sustinent F^1, *ut coni. Madvig*^{2}

Ψ's *est* after *fluidumque* may well be right, though not essential. At the end we must follow either Shackleton Bailey, supplying a participle with *aquis*, or Watt's simpler *aquas sustinent*, 'hold out against the floods' (or one might adopt Madvig's *sustinent* and alter *laeta* to something like *inundata* or *uicta*—but intransitive or elliptical *sustinent* seems harsh).

What precedes is problematic. Is *nec gramina aut pabula laeta* acceptable, with *nec* coordinate with *aut*? For *nec ... aut ...* cf. *TLL* 2.1567.84-1568.11: all the examples are from poetry, e.g. Prop. 2.28.57 *nec forma aeternum aut cuiquam est fortuna perennis*; Lucan 1.286 *nunc neque te ... pompa triumphi excipit aut sacras poscunt Capitolia laurus*. The nearest in S. is *Dial.* 6.25.3 *non illos interfusa maria discludunt nec altitudo montium aut inuiae ualles aut incertarum uada Syrtium*, but that is not parallel, for *nec* answers to *non*, and *altitudo ... Syrtium* is subordinate to the *nec*. So Watt is probably right that *nec* means 'not even', though one might consider transposing *iam nec*, thus: *... fluidumque est, nec iam gramina aut ...*, making *nec* connective, instead of coordinate with *aut* (cf. e.g. *Dial.* 2.3.5 *quomodo quorundam lapidum inexpugnabilis ferro duritia est nec secari adamas aut caedi uel deteri potest ...*; 5.2.2, 6.9.3, etc.).

Z's *pabulum aut* may just be an aberration—or could it point to the truth? Maybe (1) the ending of *pabulum* is wrong and *aut* is in the wrong place, e.g. we should read *iam nec* [or *nec iam*] *aut gramina aut pabula laeta*; but there is no strong contrast between *gramina* and *pabula laeta*; (2) could *-m aut* conceal the missing participle? but that would be the wrong position; (3) should we supply something like *gramina aut pabulum aut <arua> laeta*? but is not this, or any other, supplement somewhat repetitive? I am inclined to Ψ's text with Shackleton Bailey's or Watt's correction.

3.27.5 fame laboratur et manus ad antiqua alimenta porrigitur: qua ilex est et quercus excutitur et quaecumque in arduis arbor commissura astricta lapidum stetit.

qua ilex est F: quia i- est H: qua lex est ρ: que ilex est υ: ilex Z: quare ilex est L^1, est *del.* L^2

Z's text, *ilex et quercus excutitur et quaecumque ...*, is much more trenchant than the fussy vulgate reading *qua ilex est et quercus*.

(Accounting for the Ψ readings is not at all straightforward, but one possibility is that a scribe copied *porrigitur et quae* (his eye having jumped from *porrigitur* to *excutitur*), then noticed his mistake and continued *ilex et quercus*; but he did not completely delete the intrusive words, which were later altered to give some kind of sense.)

3.27.7 ... tandem in maiora uiolentus aberrat ...

uiolentius TX

The comparative is appropriate and gives a better rhythm than the dactylic one of the paradosis.

3.27.10 interim permanent imbres, fit caelum grauius †ac si diu† malum ex malo colligit.

ac si (sic L^2R) diu $L^2θρW^1$: ac siderum U: adsidui Z: ac sic dum *Kroll*[1]
malum Ψ: causas mali Z

Z's *causas* should not be accepted, because it spoils the brevity of the proverbial expression (Otto 207, s. *malum* 1), e.g. Liu. 7.4.7 *malum malo augere*, S. *Tro.* 427 *exoritur aliquod maius ex magno malo.* Editors print *ac sic diu*, but that cannot be right, for *sic* has no point, and it is not the paradosis anyway. Read: ... *grauius, adsidue malum ex malo colligit, ...* (for *adsidue* in the context of natural phenomena cf. 3.10.2, 4a.2.29, 6.31.1, 7.29.3).

3.27.14

exspatiata ruunt per apertos flumina campos,
cumque satis arbusta simul pecudesque uirosque
tectaque cumque suis rapiunt penetralia templis.
si qua domus mansit, culmen tamen altior huius
unda tegit, pressaeque labant sub gurgite turres.

magnifice haec, si non curauerit quid oues et lupi faciant.

2 cumque - 5 tegit Z: *om.* Ψ 3 templis Z: sacris *Ouid.* 4 mansit culmen Z: mansit potuitque resistere tanto / indeiecta malo, culmen *Ouid.*

Oltramare in his article (Oltramare[1] 35) and edition (152 n.1) gives strong reasons for adopting Z's longer form of the quotation from Ovid

Met. 1.285-90, but then prints the shorter version in his text. Z's *templis* and the omission of a line (not noted by Oltramare), with no detriment to intelligibility, suggest this is S.'s quotation from memory (he commonly misremembers or adapts quotations), not lines added by a scribe from manuscripts of Ovid (from which I have not seen these variants recorded). Could S. have said *magnifice haec* about Ψ's truncated version?

3.28.1 faciet pluuia segetes malas, fructum grando decutiet, intumescent riuis flumina, sed resident.

S. is arguing that normal forms of severe weather are not sufficient to cause the great flood. In the context *riuis* will not do; J. Müller[3] suggested *nimbis*, Watt *riui et* (but we really want a cause of the waters' rising to balance *pluuia* and *grando*). Another possibility is *niuibus*, referring to melting snow (cf. 4a.2.19-20), which complements *pluuia* and *grando*. It may be objected that not all rivers are swollen by melting snow; but equally not all crops are ruined by rain or hail.

3.28.2 ubi instat illa pernicies, mutarique humanum genus placuit, fluere adsiduos imbres et non esse pluuiis modum concesserim, suppressis aquilonibus et flatu sicciore austris nubes et amnes abundare. sed adhuc in damna profectum est.

instat Ψ: nutat Z pluuiis modum Z: m- p- Ψ concesserim Ψ: -int Z et flatu Ψ: efflatis Z

austris is problematic: it must be causal ablative with *abundare*, but (a) its juxtaposition with the other ablatives is awkward, (b) ablative of cause with *abundare* is unparalleled; it elsewhere takes ablative copiae (Axelson[1] 57 n. 12). Gemoll[2]'s *austri* will not do because, as Axelson says, *auster* is a moist rainy wind, and is not characterised by *flatu sicciore*. Maybe one should supply something like ... *sicciore, austris<que excitatis* (or *concitatis* or *inmissis*)> *nubes et amnes abundare*. Other supplements are possible.

3.28.5 si quis excelsa perlibret, maria paria sunt: nam par undique sibi ipsa tellus est (caua eius et plana eius exiguo inferiora sunt, †sed istis† adeo in rotundum orbis aequatus est), in parte autem eius et maria sunt, quae in unius aequalitatem pilae coeunt.

planis *Gercke in app.*: plano *Alexander* eius[2] *om.* F: editis *Schultess*[2] exiguo Z: *om.* Ψ sed istis adeo (a deo R) Ω: set <ex> istis a deo *Rossbach*[3] *942*: sed totus a deo *Schultess*[2]: sed ipsi alueo *Gercke in app.*: sed iustus a deo *uel* sed istis <inter se digestis> a deo *Busche*

Z's *exiguo* should be accepted, for why would it have been added? Even if a supplement were felt to be necessary, a glossator would probably not have chosen such a relatively rare expression (cf. *TLL* 5.2.1478.68-70). Several scholars have attempted to provide an ablative of comparison with *inferiora*: Professor Nisbet convincingly suggests changing ... *sunt, sed istis adeo* ... to ... *sunt editis: adeo* ... Other attempts to emend *sed istis* are singularly unconvincing.

The double *eius* is perhaps just tolerable (though Rossbach[6] should not have defended it with the pointed anaphora of *ex his* in §8), but if one has to be deleted, maybe it had better be the first.

3.28.7 aqua et ignis terrenis dominantur, ex his ortus, ex his interitus est.

terrenis Ψ: terrena Z

If Ψ is correct then Z's reading is not readily explained. Maybe S. wrote *<in> terrena*, and the *in* was lost after *-is*. For the construction cf. *Dial.* 4.36.6 *nullus adfectus est in quem non ira dominetur.*

3.29.5 ergo quandoque erit terminus rebus humanis, cum partes eius [sc. terrae] interire debuerint, aboleri<q>ue funditus totae, ut de integro totae rudes innoxiaeque generentur nec supersit in deteriora praeceptor, plus umoris quam semper fuit fiet.

abolerique Cam., *ut coni. Castiglioni*[2] *65*: aboleriue Ω tot(a)e rudes innoxi(a)eque Ψ: rudes innoxiique Z

Z's text is preferable, because only human beings can properly be *innoxii* and have a *praeceptor*. In Ψ *totae* was erroneously repeated, and then the gender of *innoxii* assimilated. A supplement like *rudes <homines>* is perhaps worth considering, though the sense is clear enough after *rebus humanis* just before.

3.29.7 ... ut quaeque proxima terris fluentibus fuerint, ipsa eluentur stillabunturque ...

eluentur Hρ: eleuentur Fυ: soluentur Z

eluo can be used of water washing away or dislodging something (*OLD* s.v. 3), cf. 6.20.6 *potest terram mouere aqua, si partes aliquas eluit et adrosit*, Plin. *Nat.* 34.157 *aqua inmissa eluente calculos nigros.* But the present passage is about earth turning to water (cf. *terris fluentibus* just before), not just being washed away, so *soluentur* is preferable.

3.30.3 mentior nisi eruentibus terram umor occurrit, et quotiens nos aut auaritia defodit aut aliqua causa penetrare altius cogit, eruendi finis aliquando est.

nos *hic* Ψ: *ante* defodit ZL^2 aut auaritia L^2, *ut coni. Mueck*: aut aut auar- Z: auar- aut Ψ: auar- auri *Schultess*[2] ali[qu]a *Gercke in app.* eruendi *Muretus*: -enti Ω: erumpendi *Leo*: <idem> eruendi *Castiglioni*[1] finis <is> *Gertz*[2] aliquando Ψ: aliquido ZL^2: aliquando <in liquido> *G. Müller 35*: aliquando <unda> *Walter*[6]

aut auaritia is correct, unless Z's reading is not the result of dittography of *aut*, but a relic of *aut auri auaritia.* But the addition of *auri* is not necessary, nor an improvement.

Alexander correctly advocates reading *eruendi finis a liquido est*, where *a* means 'because of, as a result of'. *aliquando* is the wrong sense, and Gertz's, Müller's and Walter's supplements are unnecessary.

3.30.4 omnium tunc mare ora fontium implebit, et maiore hiatu soluet. quemadmodum corpora nostra ad egestum uenter exhaurit, quemadmodum in sudorem eunt uires, ita tellus liquefiet ...

ad egestum Ψ: a deiectu Z

egestus could mean 'excretion' (so *egestio*, *TLL.* 5.2.248.9-44), but what does *ad* mean? Also, the comparison is with earth liquefying, so we want a reference not to ordinary excretion, but to diarrhoea, for which the ordinary word is *deiectio* (*TLL* 5.1.401.76-402.4, including S. *Ep.* 120.16). *deiectu*, though not found elsewhere in this sense, must be right (cf. Bourgery 263-4 on S.'s neologisms in *-us*). So read *[a] deiectu* (the *a* is dittography from *nostra*).

One might consider *<re>soluet* to improve the rhythm, but *soluet* gives an acceptable clausula.

3.30.8 omne ex integro animal generabitur ...

generabitur ς, *Fortunatus*: gloriabitur Ψ: orietur Z

Z's reading gives a suspiciously long string of short syllables, and looks like a conjecture. *gn̄abitur* was read as *gl(ori)abitur.*

4

Notes on *Naturales Quaestiones* Book 4a

4a.pr.6 Crispus Passienus, quo ego nil cognoui subtilius in omnibus quidem rebus, maxime in distinguendis et curandis uitiis ...

curandis Ψ: enarrandis Z

Oltramare[1] 19 says 'Il est certain que Sénèque attribue à son ami Passienus une grande finesse de psychologue, mais non pas les dons d'un médecin des âmes', which is quite right, for the quality in question is his *subtilitas.* Yet Oltramare then goes on to reject Z's *enarrandis*, saying we should expect *describere* or *notare*, not *enarrare.* But cf. 2.52.1 *pauca adhuc adiciam ad enarrandam uim fulminis*; 6.8.3 ... *opiniones, quibus enarratur Nili aestiua inundatio* ...; *Ep.* 121.11; *TLL* 5.2.551.47-552.2. So Z is right.

4a.pr.7 ... quemadmodum non dubiam fortunam maris, non emendi uendendique litem subeant, non incertam fidem ruris ...

litem Ψ: latam Z: aleam *Muretus*

litem, the quarrelling involved in buying and selling, is possible, but the context is rather about the risks of such ventures than the quarrelsomeness. So Muretus' *aleam* is better sense, and improves the rhythm. Z's *latam* is nearer the truth than *litem.* Cf. Pompon. *dig.* 18.1.8.1 *et sine re uenditio intelligitur, ueluti cum quasi alea emitur, quod fit, cum captum piscium ... emitur*; Paul. *dig.* 18.4.7. For *aleam subire* cf. *TLL* 1.1522.26-9, including S. *Ep.* 117.20; for *litem subire*, *TLL* 7.2.1497.45-7.

4a.pr.8 'te' inquit 'longiorem Fido Annaeo iurabo et Apollonio pycte, quamuis staturam habeas †thecisum† Thraece compositi.

thecisum *uel* thetisum *uel sim. codd.*: Thraecis cum *Lipsius*[1] *2.9*: pitheci cum *Kroll*[1]: nani cum *Friedländer 327*: pithecii cum *Alexander*

With Lipsius' emendation, the point must be not that Thracian gladiators were small (there is no evidence of this), but that they are hunched behind their smallish shields—*parmae*—when fighting. In favour of *pitheci*, there is evidence of monkeys fighting in the arena (as Warmington says, it does not matter what sort of primate animal is meant here): cf. Juv. 5.153-5 *tu scabie frueris mali, quod in aggere rodit / qui tegitur parma et galea metuensque flagelli / discit ab hirsuta iaculum torquere capella* (and Courtney), Mart. 14.202.1 *callidus emissas eludere simius hastas.* There is no evidence of monkeys being paired with Thracians, which for Warmington is an argument against *pitheci.* But of course the point is the contrast in size and the humorous improbability of the pairing.

4a.pr.13 cum quis ad te adulator accesserit, dicito: ...

laudator Z

Parroni[2] supports *laudator* on the grounds that, in what follows, *laudo* and related words occur at §§14, 18, whereas in what precedes *adulor* and related words are used at §§3, 5, 6, 9, and *laudare* only at §10: so a scribe is likely to have remembered previous occurrences of *adul-* and changed *laud-*, rather than the other way round. But there is a difference in nuance between the two words: *laus* is neutral, and can be objective and deserved, whereas *adulatio* is by definition extreme and often insincere. Here the speech that follows shows that *adulator* is more appropriate.

4a.pr.14 quamquam paupertas alia suaderet, et ingenium eo duceret ubi praesens studî pretium est, ad gratuita carmina deflexi me, et ad salutare philosophiae contuli studium.

carmina deflexi (flexi F[1]: deflexit ρ) me Ψ: deflexi ad carmina. me Z

Z's text should be repunctuated thus: ... *ad gratuita deflexi: ad carmina me et ad salutare philosophiae contuli studium...* This is preferable (a) because *me* is now in its correct Wackernagelian position, (b) on grounds of rhythm (*gratuītă dēflēxī*), and (c) because *gratuita* no longer describes *carmina* alone, as it does with Ψ's text (for *philosophiae studium* is equally *gratuitum*—it does not earn anything). This means that *deflexi* is intransitive: S. uses intransitive *flecto* at S. *Nat.* 7.21.3, 7.28.1, *Dial.* 9.9.3, and for intransitive *deflecto* cf. *TLL* 5.1.357.49-65 (including Plin. *Ep.*, Tac., Suet.).

4a.1.2 reliquo uero aestatis (Danuuius) minuitur, et ad hibernam magnitudinem redit atque ex ea demittitur.

ex ea demittitur (dim- FP²RW) L²Ψ: ex eadem it Z

This is usually understood to mean that the river falls to its winter level and then sinks lower still. But this is the only case of *demittor* = 'fall in level' in *OLD* s.v. 5c, and *TLL* 5.1.491.84-492.10 shows that *demittor* of rivers etc. elsewhere means *defluo*. Furthermore, is *ex* the most natural preposition if this were the sense? In view of L²'s reading one should perhaps not set much store by Z's, but it might suggest that *demittitur* is a stop-gap. *deminuitur* would give appropriate sense, but with *minuitur* just before it seems unlikely, unless there could be a contrast in sense between the compound verb and the simple. Conceivably there is a lacuna, *ex ea demit* <***> or *ex ea demit*<*ur* ***>, though I cannot see what might be missing; or maybe *demittitur* is a corruption of *deminuitur*, but that in turn was a corruption caused by iteration of *minuitur*, concealing something like *decrescit* (cf. 3.28.6, 6.7.4). Alternatively, as Professor Nisbet suggests, *ex ea* could be a corruption of some past participle, meaning '(flows down) in a diminished state' (*extenuata* does not seem right). So *demittitur* should be printed, but without total confidence.

4a.2.1 ... tantum usura quantum ...

Unlike modern editors I am reluctant to believe that S. wrote *tantum usura* (cf. H-Sz 123 and K-S 1.383 for *utor* with accusative in early and late Latin). *hausura* (E² uar. lect., Fortunatus) is an easy and appropriate correction.

4a.2.1-2 ... et qui insuetam aquis caelestibus terram non adiuuent. unam ut scis Aegyptus in hoc spem suam habet.

Thus Ψ, but for the latter clause Z has *unam ut scis in hoc aegiptus spes suas habet.* Z's *spes suas* gives a better clausula, and for the plural cf. e.g. *Ben.* 7.31.5 *adeo ad bonas spes pertinax animus est.* But then what is to be done with *unam*? One might in any case think of *uno* (cf. Lucan. 8.447 *in solo tanta est fiducia Nilo*). But how, without *spem*, did Z too arrive at *unam*? Either by dittography of *ut* creating the *m*, or from *terram* near the end of the previous sentence.

4a.2.2 proinde aut sterilis annus aut fertilis est, prout ille magnus influxit aut parcior.

sterilis ... fertilis Ψ: f- ... s- Z   magnus Ψ: largus Z

I suspect that Z's *fertilis ... sterilis* is preferable because it is better to have the pairs of adjectives after *proinde* and *prout* in corresponding, rather than chiastic, order. *magnus* is acceptable (3.16.1, *TLL* 8.124.9-14), but *largus*, 'generous', is more colourful and contrasts more pointedly with *parcior* (for *largus* of rivers and springs cf. Lucan. 9.608, Plin. *Nat.* 2.232, Frontin. *Aqu.* 73.6).

4a.2.5 ibi per arduas excisasque pluribus locis rupes Nilus insurgit.

abscisasque Z

excisas rupes should mean 'rocks cut out, dug out' (usually of quarrying or mining when applied to rocks: *TLL* 5.2.1241.43-53, 1242.59-84); even if it could mean 'eaten away, pitted' (and there is no evidence of it being applied to the action of water), why is that relevant here? *abscisas* means 'sheer, precipitous' (*OLD* s.v. 1, *TLL* 1.150.39-52, S. *Dial.* 2.1.2, *Ep.* 70.21), which is obviously right.

4a.2.7 primum incrementum Nili circa insulam, quam modo rettuli, Philas uisitur. exiguo ab hac spatio petra diuiditur.

Translators generally assume that *petra* is ablative and the subject is *Nilus* understood. More probably *petra* is subject and the meaning is 'a short distance away from this (island) is a crag' (see *TLL* 5.1.1600.59-

1601.36 for this use of *diuido*); otherwise the interlacing word order (*exiguo* with *spatio*, *hac* with *petra*) is unusual for prose.

4a.2.10 adeo nihil exedit nec abradit

exedit HρU: exesit Z: excedit FW

exedit must be present tense, which is highly unusual (*OLD* s. *exedo* cites this passage and Colum. 8.8.12 *exedere* (s.v.l.)—there the variant *exerere* should be adopted; *TLL* 5.2.1315.19-21 adds Ser. Samm. 89 and other late writers). Z's reading points us to *exes[i]t*, the usual classical form (S. *Ep*. 79.2; *TLL* ib. 17-19).

4a.2.15 *plerique pereunt*: translators are right that *plerique* here means 'many', as also at *Ben*. 2.4.1 and *Dial*. 9.7.5. (*OLD* s.v. *plerusque* 4 does not recognise this meaning in S.)

4a.2.18 ... nullum materiae superadornatae manet operimentum Ψ

... nulli materiae superne adornatae manet operimentum Z

nullum could easily have been attracted to *materiae*, but corruption could equally have gone in the other direction with repetition of the *m* of *materiae*; *nulli* is the more likely sense. *superadorno* is a hapax, and should perhaps be printed as two words if it is correct; but Z's *superne* is just as good.

4a.2.20 nec Rhenus nec Rhodanus nec Hister nec Caystrus †subiacent molo† aestate proueniunt; et illis altissimae ut in †septemtrionibus iugis† sunt niues.

subiacent molo ZρW: subiacent. in sola U: subiacens Tmolo *Gercke[1] 94*: <qui montibus> subiacent Nili modo *Shackleton Bailey* illis Ω: illic DX, *ut coni. Vassis* septemtrionalibus *Erasmus[1]* iugis ZL²U: iuges W: iugiter ρ

Pace Shackleton Bailey, mention of mountains is not imperative in the first part of the sentence, for §19 has established that rivers close to mountains are in question; and in any case in the next part of the sentence

Erasmus's *in septembrion<al>ibus iugis* supplies a reference to mountains, and is closer to the paradosis than the modern vulgate, *in septemtrionibus iugiter sunt niues.* (Add that S. nowhere else uses *iugiter* or the adjective *iugis*, though of course they are in any case rare in classical Latin.) So Gercke's *subiacens Tmolo*, a tiny change, is fine: it establishes that, like the famous rivers just mentioned, the Cayster too flows from a mountain. (Koeler ad loc. shows that ancient geographical writers discussed the flooding of the Cayster.)

Oltramare rightly observed that the Cayster, in Asia Minor, is scarcely in the north, so either there is corruption or S. has made a mistake. The problem lies in the second part of the sentence. (a) If *illis* refers to the rivers, then *ut in septemtrionalibus iugis* is a strikingly compendious expression (for 'as is to be expected in the case of rivers that rise in northern mountains'); also (b) if *ut* ... means 'as is natural in northern mountains/the north' (which is how it is usually taken), then it cannot apply to Cayster, unless S. has made a mistake; and if it means 'just as in ...' it can properly apply only to the Cayster. Vassis's *illic* helps with (a) (it refers to the region, not the rivers), and helps with (b) too if *illic* refers to the region of the Cayster alone, the last of the preceding trio. Another possibility is *illi*, referring to *Tmolo*. The mountain is today 2157m high, so presumably it does have snow in winter.

Professors Nisbet and Watt independently suggest *suppetunt* instead of *sunt*.

4a.2.23 praeterea ex mari purus et caeruleus efflueret, non ut nunc turbidus uenit.

uenit Ω: ueniret *Fortunati cod.*: ueniet *Skutsch*

The last phrase of the paradosis could be translated '... not in the way that it now comes, turbid', but recent editions understandably adopt *ueniret.* However, the triple trochaic caesura is not ideal. Skutsch's *ueniet* is scarcely possible sense. Could the verb have been interpolated to fill out S.'s ellipse?

4a.2.29 ob hoc Pontus in inferum mare adsidue fluit rapidus ...

inferum ZU: infernum ρ: *quid* W[1], *incertum*: finitimum *Watt*

Editors print *infernum*, but *inferum* is preferable, as better attested and more frequent with *mare* (see below). (Add that elsewhere in S. *infernus* never means simply 'lower', but is used of things underground or underworldly.) However, will either word do? For *mare inferum* is a standard expression for the *mare Tuscum* (*TLL* 7.1.1389.38-50), and *mare infernum* is a rare, initially poetic, alternative (*TLL* 7.1.1371.6-9, Lucan. 2.400, Sidon. *carm.* 5.442, Cassiod. *var.* 12.14.4, apart from our passage). The context could make it clear that 'lower sea' here exceptionally refers to the Aegean or Mediterranean (so with *TLL* 7.1.1371.7, rather than the Propontis, with *OLD* s.v. *infernus* 1d); but Watt suggests *finitimum*, and I wonder if S. wrote *in nostrum mare.*

5

Notes on *Naturales Quaestiones* Book 4b

4b.3.5 potest (grando) ... corrotundari et totiens per spatium iacens aëris densi deuoluta, aequabiliter atque in orbem teri.

iacens Z: *om.* Ψ   aeris densi ZΘPW: d- a- δRU

iacens hardly looks like an interpolation. It means 'motionless, still, calm' (cf. *OLD* s. *iaceo* 8; *TLL* 7.1.24.7-36, esp. 37-44); cf. 6.20.7 *palustres et iacentes aquas*; 2.6.4 *hic (sc. spiritus) mare per se languidum et iacens incitat*; 6.18.1 *(spiritus) iacet innoxius*; 6.27.3 *aër ... qui ... inter paludes iacet*; 7.22.1. *spatium* means a 'stretch', almost a 'layer' of dense air; cf. 1.14.1 *sunt chasmata cum aliquod spatium caeli desedit*; 2.10.1 *sed (sc. aër) non per omne spatium sui similis est.* As the hail descends from a great height it repeatedly (*totiens*) passes through a still layer of dense air, which wears it away evenly (*aequabiliter*) until its shape is spherical.

4b.3.5 ... et (nix) non per magnam altitudinem cadit, sed circa terras initium eius est. ita ne longus ille per aera sed ex proximo lapsus est.

ne ZΘρU: non δW: nec STX, *ut coni. Larisch*[1] *51*   longus *Gertz*[2]: longus quidem Z: longius Ψ

ne cannot be right, and Z's *ne ... quidem* gives a wrong emphasis, so *ne* must be changed to *non* (less likely to *nec*, unless there is a lacuna, *nec* <*... nec*> *longus*). *non longus quidem* is possible: cf. e. g. in Cicero, *Cael.* 71 *non exstinctam illam quidem sed repressam*, *Phil.* 1.20 *non centurioni quidem solum sed equiti etiam Romano*, *de Orat.* 3.15 *non a te quidem ... sed a ceteris*. However, *non ... quidem sed ...* is not found in S., and perhaps he would have written *non quidem longus*: cf. *Dial.* 4.10.3 *non quidem obtusa nec tarda, sed acumine nostro male utentia*, 5.38.1 *'non quidem' inquit 'irascor, sed dubito tamen ...*; 11.8.1, 12.17.3, etc. However, Z's *quidem* could in any case be an interpolation following on the corruption of *non* to *ne*, so it is best to follow δ.

4b.3.6 grando nihil aliud est quam suspensa glacies, nix [in] pruina pendens [congelatio]. illud enim iam diximus, quod inter rorem et aquam interest, hoc inter pruinam et glaciem, nec non inter niuem et grandinem interesse.

in *et* congelatio *secl. Madvig*[2] rorem et aquam Z: aquam et rorem Ψ: rorem et pluuiam *Gercke in app.* grandinem $ZL^2\pi$: glaciem θ

The above text has been printed by Gercke and later editors. The last sentence refers back to the first half of the book that is now lost, but the point is usually taken to be that dew, frost and snow all have a large admixture of air, whereas water, ice and hail contain no air (cf. 4b.13.2 *cum quaerimus quomodo nix fiat et dicimus illam pruinae similem habere naturam, plus illi spiritus quam aquae inesse* ...). Gross 195-6, however, thinks that the contrast is between phenomena formed on the ground and those formed in the atmosphere: dew is formed on the ground, rainwater (*aqua*) in the atmosphere, and to continue the same contrast he proposes to read ... *hoc inter pruinam et niuem nec non inter glaciem et grandinem interesse*. But this rewriting is unnecessary: both contrasts are present in the theories alluded to—with/without admixture of air; formed on the ground/in the atmosphere—and Gross gives no reason why the last sentence must refer to the latter contrast rather than the former.

4b.4.1 nam ut fatear tibi uerum, mihi quidem <facile> persuadetur, qui me usque ad mendacia haec leuiora, in quibus os percidi, non oculi erui solent, credulum praesto.

fatear k^2, *ut coni. Skutsch*: fallar Ω facile *suppl. Shackleton Bailey* percidi] perlini *Shackleton Bailey*

The meaning is: '... these less serious lies, for which you usually get your face slapped, not your eyes gouged out ...' For *os percidere* see Plaut. *Cas.* 404 *percide os tu illi odio*, Mart. 2.72.3-4 *os tibi percisum quanto non ipse Latinus / uilia Panniculi percutit ora sono*. Shackleton Bailey's *facile* is attractive, but his wish to see a reference to two proverbs (*cornici oculum* and *os sublini*) is unnecessary.

4b.6.2 protinus illae nubes alio declinabant, cum aliquid gustassent sanguinis.

gustassent Ψ: gustauerant Z

Z's *gustauerant* givs a better clausula, and a rare pluperfect indicative with *cum* describing repeated action. The imperfect and pluperfect indicative were dying out in such clauses in the 1st cent. A.D., see H-Sz 624. In S. cf. *Nat.* 1.16.4 *cum caput merserat inguinibusque alienis obhaeserat ...*, *Ep.* 108.14 *cum coeperat ...*, but at *Nat.* 4b.5.6 read *ita quae* [Winterbottom: *itaque* Z: *itaque cum* Ψ] *pluuia futura erat, grando fit ...*

4b.7.1 alteri suspicari ipsos aiunt esse in ipso sanguine uim quandam ...

ipsos Zθπ: se δ

ipsos is the paradosis, and *ipsum* for *se* is found in S. (see below), but here it seems undesirable with *ipso* four words later. One could alter to *se* with δ (though how was *se* altered? by anticipation of *ipso*?), but instead one could omit it altogether (with improvement of the clausula, *sūspĭcā-r(i) [ipsos] āiūnt*). Cf. *Ep.* 1.4 *non possum dicere nihil perdere,* and Reynolds' app. crit. ad loc., Bourgery 357-8. (For *ipsum* = *se* in S. cf. *Nat.* 6.5.1 *quidam liquere ipsis ... dixerunt*; 7.9.2 *plurium uentorum inter ipsos luctatio*; *Dial.* 10.2.5, 10.20.1, 11.17.2, *Ep.* 21.9, 98.10.)

4b.8 aiunt niuem in ea parte aëris fieri quae prope terras est.

prope terras est Vθπ: est prope terram (-as A[c]) A[1]B: prope a terris est Z

Axelson[2] 50 n. 32 observes that *prope* + accus. is not found elsewhere in S., so Z is preferable. One wonders why S. did not write *quae prope a terra est* for the clausula, but that is not sufficient reason to change the text.

4b.10 ideo minus calefit.

cal[e]fit will give a better clausula. The forms are variants e.g. at 1.2.10 *calfactus* Zθ: *calefactus* απ; 3.24.3 *calfacit* ZV[c]: *calefacit* Ψ.

4b.11.3 ... in lusoria pila: non multum illi commissurae et rimae earum nocent, quominus par sibi ab omni parte dicatur.

earum ZδθPυ: *om.* R, *del. Vottero*[3]: corii *Cornelissen*

Oltramare, Corcoran and Codoñer Merino do not translate *earum*, although like earlier editors they print it; and indeed it can only refer to *commissurae*, which gives nonsense; *illi* just before shows that we cannot be thinking of *pilae* in the plural. We should not omit *earum* (why was it added?), but should adopt Cornelissen's *corii* (→ *eoru* → *earum*).

4b.11.3-4 ... editi montes, quorum altitudo totius mundi collatione consumitur. qui dicit altiorem montem ...

altitudo Ψ: magnitudo Z

There is not much to choose, since both words are used in §1; but (a) *altitudo* is the more obvious word for mountains, and so could have ousted the rarer *magnitudo* (for this usage see *TLL* 8.118.27-30); (b) we are clearly comparing the size, not the height, of the *mundus*; so Z is preferable. (Palaeographical arguments are more evenly balanced: *altiorem* just afterwards could have encouraged normalisation of *magnitudo*; but duplication of the *m* of *quorum* could have led to corruption of an original *altitudo*.)

4b.13.5 unde ad hoc peruentum sit ut nulla nobis aqua satis frigida uideretur quae flueret dicam.

ad hoc Ψ: hoc Z

Z's *hoc* (= *huc*) should be accepted (cf. e.g. 2.11.2, 5.1.1, 6.14.4). The clause *quae flueret* recalls §4 *hoc quod illa (sc. natura) fluere et patere omnibus uoluit, cuius haustum uitae publicum fecit, hoc quod ... profudit.* However, we might have expected *quae flueret* to be next to *aqua*, rather than after *satis frigida uideretur* (where it could be taken for a consecutive clause), and it is not clear whether it means *quae quidem flueret* or *dummodo flueret*. Nor is it clear why the clause was added by S.: for there is no real risk of *aqua* in this sentence being thought to include ice (*TLL* 2.347.73-77 and 351.11-28 shows that *aqua* refers to ice only when accompanied by some unambiguous adjective, verb or participle; Mart. 4.18.8 is not really a counter-example, for the first seven lines show that *aquae* refers to an icicle); possibly *quae flueret* distinguishes running water (especially mountain streams and cool springs) from standing water, but in that case one might have expected S. to say 'that no water, not even running water, seems cold enough'. One wonders whether it could be a pedantic interpolation (by someone wanting to emphasise that S. is not talking about frozen water), or whether if should be changed to something like *quae <sitim re>ficeret* or (as suggested by Professor Watt) *quae <sitim> finiret* (cf. Hor. *Ep.* 2.2.146 *si tibi nulla sitim finiret copia lymphae*).

4b.13.5 itaque non aestate tantum sed media hieme niuem causa pari bibunt.

sed Zθπ: sed et δ   causa pari *Oltramare*: causa pati Z: hac (de *add.* B) Ψ

Knecht 388 says that *causa pari* is not first century A.D. Latin, and S. would have written *eadem ex causa.* The expression *causa par* is common in the sense 'an equally good case', but is found meaning 'a similar reason, the same reason' at Hyg. *Astron.* 1.5 *exortus est enim subita quaedam species obiecta nostro conspectui; occasus autem pari de causa ut erepta ab oculis uisa*; cf. also Tac. *Hist.* 1.72 *par ... exultatio disparibus causis consecuta.*

6

Notes on *Naturales Quaestiones* Book 5

5.5.1 quid ergo? hanc solam esse causam uenti existimo, aquarum terrarumque euaporationes; ex his grauitatem aëris fieri, deinde solui †impetum†, cum quae densa steterant, ut est necesse, extenuata nituntur in ampliorem locum? ego uero et hanc iudico ...

fieri δ: *om.* Ζθπ impetum (-ui A^1: in impetum B) Ω: impetu *ed. Ven.*[1]

Gercke assigns all from *hanc* onwards to an interlocutor, but there is no clear support for this in the text, and *quid ergo?* is usually followed by another question; so more recent editors are right to give the whole section to S., but they should have put a question mark, not a semi-colon, after *euaporationes.* δ's supplement *fieri* cannot be bettered. *solui impetum* is scarcely tenable: Oltramare translates 'puis perdent leur élan', Corcoran translates 'there is set free a rush of air': but *impetus* is too violent a word for the motion started by a concentration of dense vapours (*quae densa steterant*), and even if *impetus* were appropriate, *solui* is not the right verb. Vottero is surely right to follow Reinhardt's suggestion (152 n. 1) *solui <solis> impetu*; cf. just above c. 4.3 *quae (corpuscula) cum coaceruata sunt, deinde extenuari sole coeperunt ...*; Plin. *Nat.* 2.114 *posse (uentos) et solis inpulsu agi.* Professor Watt suggests supplying *solis* before *solui*, which is palaeographically simpler.

5.6 habet ergo aliquam uim <ui>talem aër, et ideo modo spissat se modo expandit et purgat, et alias contrahit diducit ac differt.

Muretus' *<alias> diducit* improves the shape of the sentence; otherwise the asyndeton between the contrasting *contrahit* and *diducit ac differt* is very awkward.

5.9.1 quare tamen tales uenti uere et aestate ualidiores sunt? ... quia uer aquosum est ex pluu<ial>ibus aquis, locisque ob umidam caeli naturam saturis et redundantibus maior euaporatio est.

pluuialibus *Gercke*: pluribus Ω

Gercke's conjecture is necessary, for *plures aquae* is not a satisfactory way of saying 'a lot of water'. Modern editions punctuate so that *ex pluu<ial>ibus aquis* goes with what precedes, and *locisque - redundantibus* with what follows; but this is awkward, because (a) initially one would naturally suppose that *locisque - redundantibus* is parallel to *pluuialibus aquis* and dependent on *ex*; only when one gets to *maior euaporatio est* is there any sign to the contrary; (b) why does S. say 'spring is watery because of rain-waters' rather than just 'because of rain'? The sense and structure of the sentence is much better if, with older editions, we punctuate after *aquosum est* and take *ex - redundantibus* with *maior euaporatio est.* It is maybe sufficient to put a colon after *aquosum est,* but it is very tempting to insert *et* before *ex*, as did Fortunatus' manuscripts, because *ex pluuialibus aquis etc.* is part of the answer to the original question, not an explanation of *uer aquosum est.*

5.10.1 hoc falsum esse ex eo apparet quod aura in omnem partem uehit et contra ortum plenis uelis nauigatur; quod non eueniret si semper uentus ferretur a sole.

ferretur Ψ: fieret ZL^2

Parroni[3] advocates *fieret*, arguing: 'In effetti il vento non può essere "portato" dal sole, ma semmai "causato", "originato" da esso. Inoltre non si vede perché Seneca avrebbe dovuto dire *ferretur a sole* invece che *ferretur sole.*' But *ferretur a sole* means rather 'came from the direction of the sun' (cf. *inde flatus est unde sol* in 5.9.5; for *ab* of the direction from which the wind comes, cf. 5.16.4 *qui surgit ab oriente aequinoctiali*, and passim in that chapter). Here *fieret* is also possible, but it gives a less good rhythm than *ferretur a sole.*

5.10.4 et ob hoc a solstitio illis (sc. etesiis) initium est ultraque ortum Caniculae non ualent quia iam multum e frigida caeli parte in hanc egestum est ac sol mutato cursu in nostram rectior tenditur et alteram partem aëris attrahit alteram uero impellit.

egestum δ: eccestum Zθρ: excessum υ mutato δHπ: -atio ZF nostram HρW: nostra ZAVFU: ara B: austrum *Madvig*[1] *121-2* tenditur Ω: tendit *Bongarsius*: intenditur *Haase* uero Ψ: *om.* Z

1. I have given this sentence without punctuation. The differing punctuation of editors and translators shows the difficulty of construing it: some put *ultraque ortum Caniculae non ualent* in parentheses, so that the following *quia*-clause explains *et ob hoc ... initium est*—but this is a clumsy sentence structure; others, more naturally, take the *quia*-clause to explain *ultraque ... ualent.* In other words, it is unclear whether the *quia*-clause accounts for the start or the end of the etesians!

2. The statement that the etesians do not last (or are not so strong) after the rise of Sirius is out of line with all the other ancient evidence, which either makes the etesians start with the rising of Sirius, or makes them start earlier at the solstice and continue well beyond Sirius' rising. (See Rehm *RE* 6.714; Arist. *Mete.* 2.5, 361b35-6 says the etesians blow μετὰ τροπὰς καὶ κυνὸς ἐπιτολήν, cf. 362a12 μετὰ τὰς θερινὰς τροπάς; calendars usually put the Πρόδρομοι (see *RE* s.v.) between the solstice and the rising of Sirius, the etesians after the rising.) If this statement were the only problem, one might think it was just S. nodding more than usual, but if he had said that the etesians stop or lose their force after the rising of Sirius, how could he go on to say in the next sentence *sic ille etesiarum flatus aestatem frangit et a mensium feruentissimorum grauitate defendit*?

3. The *quia*-clause has its own problems. Madvig saw that *sol mutato cursu* means the sun has passed the solstice and has begun to travel south along the ecliptic, away from us. *tenditur* cannot be right. Pincianus gives possible sense with *nostra regione cedit* (for *in nostram rectior tenditur*), but the conjecture is too wayward. With Haase's conjecture, *in nostra rectior intenditur*, the meaning is 'the sun is directed more vertically at our region', but this is false: after the solstice the sun is moving south, away from our region, lower in the sky and further from the vertical. (S. has made a serious error if he thinks otherwise.) Bongarsius' conjecture, *tendit*, meaning 'move' (*OLD* s.v. 8a), would give the same falsehood. However, *tendit* combined with Madvig's *austrum* for *nostra(m)* is probably right, though his *retro* for *rectior* is not necessary: as the sun moves south it is higher, or nearer to being vertically overhead, in the southern skies. (One should note that S. never uses

auster for 'south', but I do not think that is an important obstacle: the nearest parallel is 1.13.3 *a parte austri.* Gertz[2]'s *in australem regionem tendit* wrecks the clausula, and *meridiem* for *nostra(m)* would be further from the paradosis.)

4. If *nostra(m) tenditur* is changed to *austrum tendit*, a fairly literal translation of the *quia*-clause would be 'because already a lot has been emitted from the cold part of the sky into this part, and the sun, reversing the direction of its course, moves more vertically towards the south, and draws to it one part of the air, while it drives on the other.' But what is this talking about? I take it to mean that since the sun is now moving south, its own motion now assists the motion of the etesians: they have already been shifting air from the cold north to the hot south, and now the sun too pushes along (*impellit*) the air to the south of it, and continues to attract (*alteram partem aëris attrahit*) air from the north as before (cf. §3). (The mechanism by which the sun is supposed to attract the air behind it remains mysterious, but that is another matter.)

5. On this interpretation the *quia*-clause presupposes the continuation, not the cessation, of the etesians. Therefore, to return to the start of the sentence, the least we must do is put *ultraque - ualent* in parenthesis, with Gercke; yet that gives an awkward sentence structure and a falsehood that S. himself proceeds to contradict. So there must be corruption in the *ultraque* clause: the sense we want is something like 'and they are still (*or* very *or* no less) strong beyond the rising of Sirius', or '... are strong even beyond ...'. *non* is the problem, and various solutions are thinkable: perhaps *etiam*, or *non <minus>*, or *<pluri>mum*, or (as Professor Watt suggests) *<etiam>nunc.*

6. Winterbottom says that Z's omission of *uero* is perhaps right, in view of the rhythm. Maybe, but there was no strong reason to interpolate it, and the emphasis is appropriate: this is a new development, the sun pushing some air south. (Professor Watt suggests *ultro.*)

5.13.2 sic uentus, quamdiu nihil obstitit, uires suas effundit, ubi aliquo promuntorio repercussus est, aut locorum coeuntium <angustiis> in canalem deuexum tenuemque collectus, saepius in se uolutatur, similemque illis quas diximus conuerti aquis facit uerticem.

aut <ui> δ: aut <arto> *Gercke in app.*: aut <situ> *Busche*: aut <obiectu> *Parroni[3] 45-6* angustiis *suppl. Skutsch*

Some supplement is needed, and Skutsch's gives the best sense. But is *collectus* acceptable? *colligi* can be used of the formation of a wind (5.8.3, where exhalations collect to form the wind; Lucr. 6.124, 558, 571, all of winds gathering their strength), but it seems less appropriate here for a wind being forced into a narrow channel. *coniectus* (*cōiectus*) is a slight change, and it is the word used in a similar context at 6.20.4 *nonnumquam (uentus) in terrena itinera coniectus, et exitum quaerens*; cf. 7.27.6.

5.13.3 †haec† fere omnia pericula uenti erupti de nubibus produnt, quibus armamenta rapiantur et totae naues in sublime tollantur.

ha(e)c (hic B^{c}) Ω: et *Axelson*[1]: ac *Vottero*[3]: hinc *cod. Nicotianus*: sic *Gercke* omnia <maritima> *uel* <marina> *Watt* de Ψ: *om.* Z produnt ZVθπ: -deunt AB: praebent *Axelson*[1]: producunt *Watt* <ut> quibus *Axelson*[1]

Axelson says that *prodo* never has the sense 'produce, cause'. *OLD* s.v. 2b does allege this sense here and in other passages: at Sil. 9.2 the meaning is 'show, reveal' (*signa per Ausoniam prodentibus irrita diuis*), but 'produce, cause' is the required sense at Apul. *Plat.* 1.18 *insaniam ... quam uitiosa qualitas corporis prodit* [*prodiit* MSS.], Nerat. *dig.* 41.1.14 *litora publica ... quae primum a natura prodita sunt*; and this sense is obviously close to *OLD* 2 '(of a race, country, etc.) To give birth to'. So these passages support *produnt* here.

Z omits *de* before *nubibus*. S. does not elsewhere use *erumpo de* (found at Manil. 4.589-90), but *erumpo* plus ablative is found at 3.7.3 and so should be preferred here.

5.14.2 nam ne haec quidem supra terras quia uidentur sunt, sed quia sunt uidentur: illic quoque nihilo minus ob id sunt quod non uidentur flumina; illic scias licet nostris paria sublabi ...

nichil minus ob id θπ: n- m- ob hoc V: nichilo m- (illic quoque *hic habet* B) ob hoc AB: nichil obit minus Z sunt (*ante* quod non) δW: sint ZθρU quod non θPU: quia non δW: que non R: non quod Z

Z's reading points to *illic quoque nihilo ob id minus sunt*, which is preferable, because the emphasis is better (*minus* goes with *sunt*—'they do not exist any the less because of this'), and because the more unusual word order was prone to normalisation. For separation of *nihilo minus*, cf. *OLD* s. v. *nihilominus* 1 fin.

The paradosis in this clause is not *sunt* but *sint*, and this should conceivably be accepted; for the subjunctive expressing a conclusion cf. 2.48.1 *omnium ergo fulmimum ... dies stata sit*; but emendation to indicative is very easy in both places.

5.15.3 a tergo lucem relinquere quae tanta spes fuit?

fuit Ω: iussit *Schultess*[2]: fecit *Watt*

The objection to *fuit* is not so much grammatical (for *spes* with infinitive cf. *Ben.* 7.26.5 ... *fastidire possessa, quae consequi posse spes non fuit*), but that the sense is wrong: for they are not hoping to leave the light behind, but are hoping to gain something by doing so. Both *iussit* and *fecit* are the right sort of sense, and the latter is palaeographically simpler.

5.17.1 Placet autem duodecim uentos esse, non quia ubique tot sint (quosdam enim inclinatio terrarum excludit), sed quia plures nusquam sunt. sic casus sex dicimus, non quia omne nomen sex recipit, sed quia nullum plures quam sex.

sint Ω: sunt L[1]O, *ut coni. Gertz*[1] *163*   terrarum inclinatio Z

Z's word order gives a better clausula (*-ātĭo ēxclūdīt*). More recent editors follow Gertz's *sunt*, but there is no need. S. uses the subjunctive in this clause to stress—in case readers are uncertain—that the rejected reason is in fact false (whereas in the next sentence no reader is going to be uncertain about the truth or falsity of the statement *omne nomen sex recipit*). In S. *non quia* is often used with the indicative even where the rejected reason is false (as in the second sentence above), but for the subjunctive cf. *Dial.* 3.6.4 *hic damnatos cum dedecore et traductione uita exigit, non quia delectetur ullius poena—procul est enim a sapiente tam inhumana feritas—sed ut ...* (here, as in our passage, a parenthesis stresses the falsity of the claim); 8.3.1 *hoc Stoicis quoque placere ostendam, non quia mihi legem dixerim nihil contra dictum Zenonis Chrysippiue committere, sed quia res ipsa patitur me ire in illorum sententiam* (without the subjunctive *dixerim* the reader might think that S. *has* adopted such a rule).

5.18.10 sic Alexander ulterior Bactris et Indis uolet, quaeretque quid sit ultra magnum mare ...

ulteriora *Gercke in app.* uolet Ω: uadet *Leo*: uoluet *Shackleton Bailey*: <noscere> uolet *Watt*

The paradosis cannot mean what recent translators want it to mean, 'Thus Alexander will want to go beyond Bactria and India'. Hence the conjectures based on *uolet*: but they all spoil an excellent clausula (*-trīs ĕt Īndīs uŏlēt*), though that could be avoided by inserting Watt's *noscere* somewhere else, e.g. after *Alexander*. However, if Gercke's *ulteriora* is accepted, then *uolet* can mean 'Alexander will want (to possess) the regions beyond Bactria and India'. For *uolo* = 'to desire to have, want (a particular thing or person)' see *OLD* s.v. 2. In S. cf. e.g. *Ep.* 95.2 *saepe aliud uolumus, aliud optamus*; 122.9 *cum instituerunt omnia contra naturae consuetudinem uelle*; *Thy.* 213 *rex uelit honesta: nemo non eadem uolet*; 246 *ego poenam uolo*; *Tro.* 967-8 *quam uellet tuos Cassandra thalamos, uellet Andromache tuos.*

7

Notes on *Naturales Quaestiones* Book 6

6.1.2 Neapolis quoque priuatim multa, nihil publice amisit, leuiter ingenti malo perstricta. uillae uero prorutae, passim sine iniuria tremuere.

nihil publice Z: p- n- Ψ perstricta Zθπ: prestr- δ prorut(a)e ZHPU: prerupte δ: prorupte FW: perrupte R: <multifariam> prorutae *Oltramare*[1] *41*

The clausula at the end of the first sentence is acceptable, but would be improved by deleting the prefix of *[per]stricta.* This would be another example of the simple verb being replaced by a compound that is more common in the required sense (see above on 1.2.9); cf. *OLD* s. *stringo* 5 for this sense.

Oltramare suggested inserting *multifariam* before *prorutae* (his edition mistakenly attributes this to Z), and some such supplement is an improvement, but there are several other possibilities, e.g. *<plurimae>*, *<nonnullae>*, *<hic illic>*, *<paucae>* (the last two suggested by Professor Watt). (Some manuscripts punctuate after *passim*, but it is less likely that the adverb follows the verb.)

6.1.5 quam latebram prospicimus, quod auxilium, si orbis ipse ruinas agitat ...

ruinas (-am W) Ψ: rimas Z

All editions read *ruinas*, and understand it to mean 'if the earth itself causes (or: threatens to cause) ruin'; *OLD* s. *agito* 11b, on the other hand, takes the phrase to mean 'collapse'. I know of no parallel for either meaning; and as for the second, while parts of the earth may collapse in an earthquake, could the whole *orbis* be said to? (At Val. Fl. 6.633 *ruinas agere* is quite different: *imber agens scopulos nemorumque operumque ruinas.*) But *rimas agere* means 'to crack, to form cracks', cf. *TLL* 1.1376.64-68, including S. *Ben.* 6.15.7 *quantum nobis praestat, qui labentem domum suscipit et agentem ex imo rimas insulam incredibili arte suspendit!* So Z's reading can be paralleled (with *ago*) and gives excellent sense.

6.1.6 ... subterraneae domus et in altum acti specus ...

in altum acti Z: in altum θπ: defossi in altum δ

δ, we may deduce, supplied the kind of participle that was missing in Ψ. Is Z's *acti* likewise conjecture? It was vulnerable to omission after *altum*, so Z may well have the truth.

6.1.11 unde scias an <non> melior eorum locorum condicio sit, in quibus iam uires suas fortuna consumpsit ...

scias ZθPW[1]: scis δ: scies U non *add. Haase*

δ has written the normal *unde scis an*, but *scias* (printed by Codoñer Merino) can be defended by analogy with *haud sciam an* (Cic. *de Or.* 1.255, 2.18, al.; Liu. 9.15.10; H-Sz 335), and *nescias an* (Hor. *Carm.* 2.4.13); also S. *Dial.* 3.1.4 *nescias utrum magis detestabile uitium sit an deforme*. (Cf. Löfstedt[1] 2.133.)

Vassis and Oltramare edn. 250 n. 2 defend the paradosis against Haase's supplement *non* on the grounds that *unde sci(a)s an* is analogous to *nescio an*, conveying a positive affirmation of what is in the *an*-clause. But this is not the case in *Dial.* 6.22.1, where the question is genuine: *unde enim scis an diutius illi expedierit uiuere, an illi hac morte consultum sit?*, 'for how do you know whether it was to his advantage to live longer, whether this death was in his interests?'; the other Senecan example is not clear-cut, *Ep.* 17.8 *tu uero, siue aliquid habes, iam philosophare (unde enim scis an iam nimis habeas?), siue nihil, hoc prius quaere quam quicquam*. So Haase's supplement should be accepted, to give the right sense in our context. (I know no examples of *unde sci(a)s an* apart from these passages of S.)

The elder Seneca does use *unde scio an* differently: *Contr.* 1.2.21 *unde scimus an cum uenientibus pro uirginitate alio libidinis genere deciderit?*, 'How do we know that she did *not* bargain ...' (Winterbottom's translation, my emphasis); so, later, does Apuleius (*Met.* 1.15 *unde autem ... scio conuectore illo tuo, cum quo sero deuorteras, iugulato fugae man-*

des praesidium?, 'How do I know you haven't slit the throat ...', 6.1, 10.26. (These are the only cases of *unde scio, etc., an* in these authors.) Compare above on 3.pr.4.

6.1.13 Tyros aliquando infamis ruinis fuit; Asia duodecim urbes simul perdidit; anno priore in Ach(a)ia et Macedonia quaecumque est ista uis mali quae incurrit nunc Campaniam laesit.

mali qu(a)e Ψ: malique Z

The above is the paradosis. It is not clear whether *nunc Campaniam* goes (i) with *incurrit* or (ii) with *laesit*, but both options cause severe problems: with (i), *laesit* has no object; with (ii), we have the nonsense *in Achaia et Macedonia ... Campaniam laesit* (unless we can take *in A. et M.* with the *quae incurrit*, a difficult word order). There are various conjectures on offer. Erasmus[1] attached everything down to *Macedonia* to the previous sentence and continued *quaecumque [est] ista uis mali [quae] incurrit, nunc Campaniam laesit,* which is not persuasive. Other conjectures may be divided according to options (i) and (ii) above:

(i) (with punctuation after *Campaniam*): Fortunatus proposed *[in] Achaiam et Macedoniam*, thus providing an object for *laesit*; Courtney changed *laesit* to *saeuiit,* which is compatible with *in Achaia et Macedonia.*

(ii) (with punctuation after *incurrit*): Gronovius[1] proposed: *anno priore in Achaiam et Macedoniam, quaecumque est ista uis mali, quae incurrit; nunc Campaniam laesit* (apparently taking *anno - Macedoniam* with *laesit*);

(iii) different approaches are: Haase transposed *quae incurrit*, thus: *anno priore in Achaiam et Macedoniam quaecumque est ista uis mali incurrit, quae nunc Campaniam laesit*; Geist changed *mali quae* to *maligne*, and Castiglioni[2] simply deleted *quae* (cf. Erasmus above), both adopting *in Achaiam et Macedoniam.*

The following points may be made about these various approaches:

1. Punctuation (i) is doubtful, because it leaves the verb *laesit* awkwardly separated from the rest of its clause by *quaecumque - Campaniam*; this applies whether we emend to *[in] Achaiam et Macedoniam ... laesit* (Fortunatus) or to *in Achaia et Macedonia ... saeuiit*

(Courtney). It is also questionable whether *incurro* can take a direct object here. The only parallel in S. is *Ben.* 1.12.1 *ingratos quoque memoria cum ipso munere incurrit*, where Gertz's *<in> ingratos* is very easy (*ingratis* vulg.). But *incurro* with direct object is found elsewhere, especially in military contexts (*TLL* 7.1.1088.78-1089.8), so it cannot be ruled out here.

2. Punctuation (ii) gives us something like Gercke's text: *anno priore [in] Achaiam et Macedoniam, quaecumque est ista uis mali, quae incurrit, nunc Campaniam laesit* (which may be how the sentence was understood by some earlier editors who printed it without internal punctuation). But *quae incurrit* is rather lame, and in any case *ista uis mali* does not really require further definition in the context.

3. I conclude that *quae incurrit* is probably not sound, since it is equally unsatisfactory whether taken with *nunc Campaniam* or on its own. We have to choose between Haase's transposition and Castiglioni's deletion of *quae* (for Geist's conjecture is implausible); and Professor Nisbet suggests that, alternatively, we transpose *incurrit* to follow *Macedoniam*. It seems to me slightly better to have Campania mentioned in a coordinate clause rather than a subordinate one, which favours Castiglioni; as he says, *quae* could be repetition from *quaecumque*, or (less likely, he thinks) added by someone who misunderstood the structure of the sentence. So read: *anno priore in Achaia<m> et Macedonia<m> quaecumque est ista uis mali [quae] incurrit, nunc Campaniam laesit*.

This supposes that *incurrit* is perfect, which is possible, cf. *TLL* 7.1.1083.6-11; here *incucurrit* would give a weaker clausula, with a series of trochees.

6.2.1 ego uero hoc ipsum solacii loco pono, et quidem ualentissimi, quoniam quidem sine remedio timor stultis est: ratio terrorem prudentibus excutit; imperitis magna fit ex desperatione securitas.

quidem <non> *Watt* stultus RUP^2: tutus *Axelson*[1]: stultis <salutis> *Shackleton Bailey*

Axelson is right that the paradosis will not do and that *stultus* is too bland, but his approach is less satisfying than Shackleton Bailey's. However, the genitive *salutis* is more Tacitean than Senecan (his parallel is *Ann.* 3.20.2 *illam obsidionem flagitii ratus*), and we might expect 'for the foolish also' or 'even for the foolish' (for the wise will certainly not feel fear). So maybe S. wrote something like *stultis <saluti>* or *<quoque saluti>* or *<(quoque) salutaris>*; or Professor Nisbet suggests changing *stultis* to *salutaris*. With Watt's *quoniam quidem <non>* ..., S. makes a general statement about fear of any kind, but it is more in keeping with the context to take *sine remedio* attributively with *timor*.

6.2.6 quid stultius quam timere nutationem terrae, aut subitos montium lapsus, ...

terrae δ: *om.* Zθπ

terrae is a necessary supplement by δ, but it is better added after *timere*, where its loss is more easily explained; this also leaves a good clausula at *nutationem*.

6.2.9 egregie Vagellius meus in illo incluto carmine, 'si cadendum est' inquit 'mihi, e caelo cecidisse uelim'. idem licet dicere: si cadendum est, cadam orbe concusso.

est inquit mihi Zθπ: e- m- i- AB: m- e- i- V: est [mihi]' inquit *Gercke* idem Ω: idem <mihi> *Gercke*: <mihi> item *Housman*[3] *168* = *Housman*[4] *1247-8*

The paradosis is as given above. The position of *mihi* after *inquit* is probably no obstacle; for *mihi* after parenthetic *inquit*, and slightly detached from what follows, cf. e.g. *Dial.* 5.23.2 *'dicite' inquit 'mihi facere quid possim ...'*; Liu. 39.28.1 *'non cum Maronitis' inquit 'mihi aut cum Eumene disceptatio est'*; Sen. *Contr.* 7.7.12 *'nihil dixit' inquit 'mihi'*. Nor should one delete or transpose *mihi* for the sake of the metre of the quotation: it is hazardous to try to impose metrical regularity on what may be a partially paraphrased quotation. So Housman's *item* is the easiest way of restoring the required sense.

6.3.3 quid ergo? non religionem incutit mentibus et quidem publice, siue deficere sol uisus est, siue luna, cuius obscuratio frequentior, aut parte sui aut tota delituit? longeque magis illa, actae in transuersum faces et caeli magna pars ardens et crinita sidera et plures solis orbes et stellae per diem uisae subitique transcursus ignium multam post se lucem trahentium.

illa is suspect, hence the *illae* of the Leipzig edition of c. 1492-5 (which is not a persuasive emendation, for why should *faces*, alone of the series, be emphasised with *illae*?); but also the lack of a verb in this sentence is awkward, because we have to understand a verb (*religionem incutiunt*) not from the *siue*-clauses that immediately precede, but from the beginning of the previous sentence. Maybe a verb has fallen out, something like *longeque magis illa <terrent>, actae ...*, or *illa <sunt terribilia>*?

6.4.1 ... ignesque nonnumquam per aliquod ignotum antea montis aut rupis foramen emittat, aliquando notos et per saecula nobiles comprimat.

comprimat Ψ: suprimat Z

Either *comprimat* or *supprimat* will give the required sense, 'check, stop' (cf. *OLD* s. *comprimo* 10c, s. *supprimo* 4b). But *comprimat* occurs five lines earlier: there is no objection to the repetition as such, but the earlier *comprimat* could have caused an original *supprimat* to be corrupted, so the latter is preferable.

6.4.2 ... temptare me uoluerim et experiri, aetas aliquid nobis aut ad scientiam aut certe ad diligentiam adiecerit.

<an> aetas AV: <si> aetas *Gercke*

We should maybe return to the vulgate *an*, which is the construction S. uses after *experiri* at *Ep.* 35.4; though *si* cannot be ruled out. In other authors, for *experiri* with *an* cf. *TLL* 5.2.1669.3-7; with *si* cf. *TLL* ibid. lines 29-34.

6.6.2 illud argumenti loco ponit, aquas esse in causa quibus hic orbis agitetur, quod in omni maiore motu erumpunt fere noui fontes (sicut in nauigiis quoque euenit ut si inclinata sunt et abierunt in latus, aquam sorbeant, quae †in omni eorum onerum quae uehit† si inmodice depressa sunt, aut superfunditur, aut certe dextra sinistraque solito magis surgit).

in omni (post quae) ZL^{2}θπ: omni δ: *lacunam post* omni *statuit Haase*: impatiens *Madvig*[2]: ui omni *Gercke*: enormi (*et* onere) *Rossbach*[6] *1484*: ui *Winterbottom*: momine *Shackleton Bailey*: nimio *Watt* (h)onerum Zθπ: (h)oneri δ: onere *ed. Ven.*[1]: onerariorum *Michaelis*[2] *345*: motu (*om.* eorum) *Leo*: pondere (*uel* <pondere> *ante* eorum) *Watt* uehunt *Shackleton Bailey*

Watt's *nimio* and Shackleton Bailey's *uehunt* are excellent, but simpler than Watt's *<pondere> eorum onerum* or *eorum pondere* is *eorum onere* of ed. Ven.[1]. For *onus* = 'weight' cf. 6.17.2 *cum (unda) onus suum sustinere non potuit*, 7.14.2 *quomodo illa moles non descendit et onere se suo frangit?*; *Dial.* 1.2.6, *Ag.* 89, *TLL* 9.2.646.37-53. Assuming that *quae* refers to *nauigia* (not to *aqua*, which would be awkward), we must understand *aqua* as the subject of *superfunditur* and *surgit;* this is possible, but maybe one could supply *<aqua> aut superfunditur ...*

6.6.3 hanc opinionem falsam esse non est diu colligendum: nam si terra aqua sustineretur et ea aliquando concuteretur, semper moueretur, nec agitari illam miraremur sed manere.

et ea (*om.* U) Ω: tota *Fortunatus* et - concuteretur *om.* U, *del. Haase* aliquando concuteretur] agitante conc- *Gertz*[2]: conc- aliquando *Skutsch* sustineretur, et *dist. Axelson*[1] concuteretur <et> *Alexander*

The paradosis is fine (cf. J. Müller[3]): the theory under attack is that the earth floats on water and so is sometimes shaken, and the objection is that on this hypothesis the earth would always be shaking. The theory comes in the protasis (*si terra - concuteretur*; I take *ea* to be ablative, like *aqua*), the objection in the apodosis (*semper moueretur*). Axelson's punctuation, making the apodosis start at *et ea aliquando*, gets the argument wrong, and the *et* is awkward, especially in view of the postulated asyndeton at *semper moueretur* (hence Alexander's insertion of *et*)

6.7.5 nec enim uideo quid prohibeat aut obstet, quominus habeat aliquod etiam in abdito litus, et per occultos aditus receptum mare, quod illic quoque ...

habeat ZH: habeat hoc Fπ: habeat habet AV: habeantur. habentur B: pateat *Watt habeat,* not the *habeatur* of editions prior to Gerkce's, is the paradosis, and as Gercke says, the subject must be understood to be *terra.* But it looks likely that Ψ had something like Fπ's *habeat hoc*; the *hoc* was omitted by H, and an abbreviation was inaccurately expanded in ABV. Possibly *hoc* stood in the archetype and was omitted by Z by accident or design. Change to *habeat haec* (i.e. *terra)* and the sentence will be improved.

6.7.6 illic quoque potest recedere ac retegere pelagus infernum; quorum neutrum fit sine motu superstantium.

recedere Ω: accedere ς, *Erasmus*[1] retegere Ψ: reicere Z: decedere *Erasmus*[1]: recedere *Fickert*: resurgere *Haase*: relegere *Gercke in app.*: surgere *Codoñer Merino*

S. is not talking about regular tide-like motions, which do not cause the sort of damage he has just described, but about subterranean tidal waves; this rules out *accedere ac recedere*, a standard phrase for tidal motion, for something exceptional is required for S. to continue *quorum neutrum fit sine motu superstantium.* There is nothing wrong with the unanimously attested *recedere* for the first verb (Vottero[3], who repeats Fickert's conjecture, compares *Ben.* 1.10.1, Cic. *Nat.* 2.132, Suet. *rell.* 157 (p. 242.11 Reiff.)). For the second, I see no verb of which Z's *reicere* and Ψ's *retegere* are plausible mechanical corruptions; a possible course of corruption is that a second verb starting with *re-* was ousted by iterated *recedere,* and Z and Ψ represent feeble attempts to produce an alternative verb. Haase's *resurgere* is possible, but perhaps better would be something like *redundare*, which clearly denotes overflowing, not just returning to the original level (cf. e.g. 3.30.6 ... *sic momento se redundantia pluribus locis maria committent*).

6.9.1 Ignem causam motus quidam †et quidam non† iudicant, in primis Anaxagoras, qui ...

Either *et quidam non* is a silly interpolation (δ omits it), or there is a lacuna. Diels suggested (ap. Gercke) *et quidam non <unicam causam>*, which is not attractive. Others, following the lead of the vulgate *et quidem non <eandem>* (Fortunati vetus lectio), suggest *et quidem nou<issimam>* or *non <primariam>* (Gercke in app.), *et quidem non*

<ignobiles> (Oltramare), *sed quidem non <uno modo>* (Codoñer Merino, but *sed quidem* is not Latin, and *inprimis Anaxagoras* is inappropriate after that). Possibly we should supply *et <ei> quidem non <*>* (*ei* was vulnerable after *et*); cf. Cic. *Luc.* 17 *id quidam e philosophis et ei quidem non mediocres*, *Fin.* 1.1 *nam quibusdam, et iis quidem non admodum indoctis*, *Sen.* 75, *Nat.* 2.29, *Off.* 1.71, Quint. *Inst.* 7.4.1. Alexander, who thinks the phrase is interpolated, points out that Oltramare would better have written *eique non <ignobiles>*. Maybe supply e.g. *et <ei> quidem* (or *eique*) *non <indocti>* (omitted by parablepsy before *iudicant*)—though the rhythm is not pleasing.

6.12.1 Archelaus †antiquitatis† diligens ait ita: 'uenti in ...

antiquitatis Ω: <inter> antiquos satis *Gertz*[2]: auctor ueritatis *Schultess*[2]: <auctor, ut est captus> antiquitatis *Garrod*[2]: <physicus, ut est captus> antiquitatis *Oltramare*: inquisit<or s>atis *Axelson*[1] *(cf. 7.3.3* diligens et ipse inquisitor*)*: <antiquus> inquisit<or s>atis *Alexander*: uir quidem satis *Vottero*[3]: inquisitor *Watt* ait ita AVθπ: ita ait Z: ait B

There is no certain solution to the crux. Parroni[4]'s defence of the paradosis, that Archelaus is called *antiquitatis diligens* because he reverted to the ideas of Anaximenes and the early Ionian physicists, is not persuasive; the point would be rather recherché for S., from whose point of view Archelaus himself belongs firmly to *antiquitas*.

Both *ait ita* and *ita ait* are extremely rare before direct speech. This is the only example in *TLL* 1.1456.14, though it is also found at Apul. *Flor.* 2, Gell. 5.15.9, Ulp. *dig.* 24.1.3.pr., Serv. *Aen.* 3.84, al. (I ignore cases where *ita* is picked up by a following *si*-clause.) Maybe B is right, and *ita* should be deleted as the product of duplication of *ait*; this would improve the clausula.

6.12.3 'quid ergo? numquam flante uento terra concussa est?' admodum raro: <raro> duo simul flauere uenti, fieri tamen et potest et solet.

suppl. Shackleton Bailey: <cum> *Gercke* <si> simul *Alexander*

Shackleton Bailey is right to supply *raro*, but *<nam raro>* or *<quia raro>* would make the argument clearer.

6.13.2 eo frigidum confluit unde uis calida discessit, et inuicem ibi calidum est unde frigus expulsum est. hoc quod dico uerum †est sed† utrumque in contrarium agi, ex hoc tibi appareat.

est sed Zδθ: est si ρU: sed W: est, scilicet *Koeler*: esse, et *Haase*

Haase may be right, but another possibility is *esse, id est*, since *utrumque in contrarium agi* is the same as *hoc quod dico* (sc. in the previous sentence), not a new fact.

6.13.3 qui [sc. calor], cum in inferiora peruenit, et eo se quantum poterat ingessit, quo densior, hoc ualidior est. †huic alius† superuenit, cui necessario congregatus ille iam et in angulum pressus loco cedit. [4] idem <e> contrario euenit: cum uis maior frigidi inlata in cauernis est, quidquid illic calidi latet frigori cedens abit in angustum ...

huic alius (h- illius U) Ω: <ubi> huic frigus (et ... cedit, idem ... *dist.*) *Gercke*: hic alii <spiritui> *Oltramare*: hic alii *Alexander*: inuicem alterius <uis> *Garrod*[2]: huic calidus <spiritus> *Codoñer Merino*: rigori cal<idi> uis *Vottero*[3]: frigori cal<idi> uis *Vottero edn.* angulum Ω: angustum ς econtrario R: contrario ZδθPυ

It is unnecessary to alter *angulum*, which gives better rhythm than *angustum* (*āngŭlūm prēssūs*).

There seem to be two stages to the two processes described by Strato:

1. in winter (1a) cold above the earth's surface drives heat below ground, and (1b) this heat drives away the cold that is already below ground, and this causes earthquakes;

2. in summer the opposite happens, (2a) heat above the earth's surface drives cold below ground, and (2b) this cold drives away the heat that is already there, and this causes earthquakes.

§4 describes the summer process clearly enough, but there is confusion in §3. As Gercke saw, *alius* can only be another bit of heat, but after *quo densior (sc. calor), hoc ualidior est* we do not want a reference to yet further heat arriving. With Gercke's conjecture and repunctuation, *<ubi> huic ... loco cedit* describes 2a. But this is awkward for several reasons: *huic* and *ille* both describe the same hot air; *cum uis maior frigidi inlata in cauernis est* repeats what is said in the previous sentence; and there is no description of 1b. It is more satisfactory to make *congregatus ille ... loco cedit* describe the reaction of the cold air already below ground to the arrival of the hot air (1b). Oltramare's and Alexander's conjectures are moving in the right direction, but lack specific reference to cold air. Vottero's earlier conjecture gives suitable sense, but the phrasing is unattractive and palaeographically the conjecture is implausible. Maybe there is a lacuna, e.g. *<frigido autem aëri qui iam sub terra collectus est> h[u]ic <c>ali<d>us superuenit, cui ... loco cedit. idem e contrario euenit: cum ...*

6.14.1 sunt qui existiment spiritu quidem et nulla alia ratione ...

et nulla Ψ: nec ulla Z

et null- is less frequent in S., and more emphatic, so Z may have altered to the more usual form. (*et null-* is found about a dozen times in S.'s prose, *nec ull-* about four dozen times, if one ignores passages where the one cannot be substituted for the other.)

6.14.2 ... spiritu ... icto uehementius et quassante uenas suas. nec, ut illi paulo ante dicebant, quibus animal placet esse terram ***. nisi hoc est, quemadmodum animal, totum uexationem parem sentiet: neque enim in nobis febris alias partes moratius impellit, sed per omnes pari aequalitate discurrit.

suas, nec (ut ... terram) <emisso> *Gercke*   *lacunam ante* nec ut *statuit Holl 21-2*   nec <haec> *J. Müller*[3]   dicebam *Madvig*[2]   placet esse Ψ: esse placet Z   *lacunam post* esse terram *statuit Reinhardt 160 n. 1*   nisi (ubi W[1]) Ω: nam si *Erasmus*[1]*:* si *Madvig*[2]: sed si *uel* at si *Castiglioni*[2]: *del. Rossbach*[5]   totum (-am F) Ω: tota ς *fort. recte*   parem sentiet Z (*et fort.* W[1]): s- p- δU: s- θπ   moratius Zθπ: moratius alias citius δ: moderatius *Muretus*: moderatius <aliis> *Madvig*[2]: morantius *Rossbach*[6] *1484*   impellit Ψ: pellit Z

nec, ut ... esse terram lacks a main verb, so there is a lacuna somewhere. Reinhardt's location is preferable, for then the missing words would have given the following *hoc* a clear sense. There are no sound grounds for altering *nisi*.

moratius = 'more slowly' is unparalleled (*TLL* 8.1501.82-4) and not very plausible, and even if it existed, it is the wrong sense, for *pari aequalitate* following shows that the comparison is of degree, not of speed. So Muretus' conjecture should be accepted.

But *alias moderatius* will not do on its own. δ's *alius citius* has dubious claim to authenticity (it is just possible that the other manuscripts omitted it independently by parablepsy), but *citius* betrays itself as a companion to *moratius*. Maybe S. wrote something like *alias moderatius* *<alias uehementius>*.

Z's *pellit* gives a better rhythm; cf. on 1.2.9 for compound verbs replacing simple ones in the manuscripts.

6.14.3 uide ergo nunc quid intret in illam spiritus ex circumfuso aëre, qui quamdiu habet exitum, sine iniuria labitur.

nunc Ω: num J^2, *Pinciani uetus lectio*: ne *Gercke, Kroll*^{1}

Codoñer Merino and Vottero return to *num*, Vottero comparing *Ep.* 9.2 *uide ergo num satius sit aut inuulnerabilem animum dicere aut animum extra omnem patientiam positum* (cf. also *Ben.* 2.18.3 *uideamus num* ...). Palaeographically *num* is slightly easier than *ne*, but *ne* could have been misread as $\overline{nc}$.

6.14.4 at si ne rimam quidem per quam efflueret inuenit, conglobatus illic furit, et hoc atque illo circumagitur, aliaque deicit, alia intercidit, cum tenuissimus idemque fortissimus et inrepat quamuis in obstructa, et quidquid intrauit ui sua diducat ac dissipet.

deicit δρU: deiecit ZθW^1 intercidit Ψ: interscidit Z: interscindit *cod. Anicianus, ut coni. Pincianus* inrepat Ω: inrumpat *Watt*

deiecit is better attested, and Z's *interscidit* is the verb S. uses in similar contexts at 3.11.1 *ruina interscindit cursum aquis*, 2.17 *dum (spiritus) luctatur per obstantia atque interscissa uadere*, 2.19; but for *intercido* cf. 1.2.9 *si ab una parte (facies) intercisa est.* (For manuscript confusion between the two verbs cf. *TLL* 7.1.2276.1-3, 2163.6-8.) But is the variation of tenses acceptable? Probably it is, for the present tenses describe continuous activity, the perfects describe one-off actions; cf. Hine[6] 326-8. But *deicit* ... *interscindit* would be easier.

6.17.1 sequitur ergo ut naturam suam exerceat, et quod semper moueri uult, aliquando et alia moueat. id quando fit? quando illi cursus interdictus est.

quando fit? quando A[1]Vθπ: tanto fit quanto A *u. l.*: tanta fit quanto B: fit ubi Z

We may take it that A[1]Vθπ give Ω's reading. They and Z give curious variants. Ψ's second *quando* offers the only case in Seneca of *quando* in a temporal clause (which is common in early Latin, and passes into romance, but is rare in later prose authors, cf. K-S 2.365, H-Sz 607), so Z's *ubi* is preferable. But what of Ψ's first *quando*? The question *id quando fit?* is rather fussy in the context, and Z's *id fit ubi* ... is preferable.

6.18.3 ita eius (sc. spiritus) non potest uis tanta cohiberi, nec uentum tenet ulla compages.

eius Ψ: est Z

eius is in a perversely emphatic position, and Z's text is much better, punctuated thus: *ita est: non potest uis tanta cohiberi ...* Here *ita est* introduces a restatement or generalisation of what has just been said. Cf. e.g. *Dial.* 11.1.1 *quid enim inmortale manus mortales fecerunt? septem illa miracula et si qua his multo mirabiliora sequentium annorum extruxit ambitio aliquando solo aequata uisentur. ita est: nihil perpetuum, pauca diuturna sunt ...*; 6.23.5; *Ep* 94.71; 97.13; 110.20; *Nat.* 3.27.2.

6.18.7 nam quamdiu sine iniuria perfluit et ex more procedit, nullus est tremor corpori: cum aliquid occurrit quod inhibeat eius officium, tunc parum potens in perferendis his quae suo uigore tendebat, deficiens concutit quidquid integer tulerat.

Schultess[2] transposed *suo uigore tendebat* and *integer tulerat*, which is supported by Alexander and accepted by Vottero. Shackleton Bailey, however, proposes *quae suum uigorem tendebant.* But what is S. talking about? About carrying heavy loads, as Shackleton Bailey seems to think? But his conjecture ruins a good clausula, and is out of keeping with the medical context: S. is talking about the effect of healthy and unhealthy *spiritus* on the body (the latter causing trembling), not about carrying heavy loads. Chapter 6 of book 2 (esp. §3 and §6) shows what *suo uigore (spiritus) tendebat* could mean: *spiritus* keeps the body in tension, according to Stoic doctrine. The clause cannot refer to that doctrine in its present position, but Schultess's transposition is very appealing, giving: *tunc parum potens in perferendis his quae integer tulerat* [the reference is to things that are potentially frightening, etc., which the healthy *spiritus* can cope with but the weakened *spiritus* cannot], *deficiens concutit quidquid suo uigore tendebat* [referring to parts of the body that now start to tremble, although they used to be held in a state of healthy tension]. (The transposition is not inconceivable: e.g. if the scribe's eye jumped from *quae* (abbreviated) to *quidquid* (ditto), then *integer tulerat deficiens concutit quidquid* could have been added in the margin and reinserted in the wrong place, at the end of the sentence; and then *integer tulerat* could have been moved to create some sense.)

6.20.1 aliqua pars terrae concaua est; in hanc aquae magna uis confluit. ex hac est aliquid tenue et ceteris liquidius. hoc, cum superueniente grauitate reiectum est, inliditur terris et illas mouet; nec enim fluctuari potest sine motu eius in quod inpingitur.

deiectum *Gercke ap. DK II.107* terris] terrenis *Gercke dubitanter ibid.*

Usually *grauitate* is translated 'a heavy mass' or something similar, this apparently being understood to refer to a solid mass (of rock?) falling onto the water. But *grauitas* nowhere else means 'a heavy weight': neither L&S nor *TLL* has our passage or this sense; *OLD* s.v. 1 finds the sense '(concr.) a heavy mass' here and in Ov. *Met.* 9.287: *tendebat grauitas uterum mihi*—but one could perfectly well translate 'heaviness' there. (*TLL* 6.2306.33-4 assigns the sense *grauiditas* to the Ovid passage uniquely.) Furthermore, *superuenio* is not an appropriate word for a solid object falling on water: it describes one thing covering or weighing down on another; see *OLD* s.v. 1, and in Sen. *Nat.* it occurs with subject *unda* (3.10.2, 6.17.2), *aestus* (6.2.5), *spiritus* (6.12.1) (7.12.4 *stella stellae superueniens* and other passages are not comparable). So here the reference is more likely to a weight of denser water covering the lighter water. But a genitive is essential: certainty is impossible, but we need something like *cum superueniente <densioris aquae> grauitate reiectum est.*

6.20.4 nonnumquam (spiritus) in terrena itinera coniectus et exitum quaerens mouet omnia. terra autem penetrabilis uentis est et spiritus subtilior est quam ut possit excludi, uehementior quam ut sustineri concitatus ac rapidus.

terra ABRW[1]: et terra ZVθP: e terra U: ut terra *Diels DK II.107* autem Ω: quoque Cam.E <nam> et terra quoque *Gertz*[2] uentis [est] *Gercke, Skutsch* et spiritus] ita spiritus *Diels ibid.* subtilior [est] B

The paradosis is *et terra autem* (which was in ed. Ven.[1]). There is no need to doubt this, cf. *TLL* 2.1593.76-1594.40; in S. cf. *Ep*. 58.15 *in rerum ... natura quaedam sunt, quaedam non sunt, et haec autem quae non sunt rerum natura complectitur ...*, 92.1 *nam illa quoque diuina ratio omnibus praeposita est, ipsa sub nullo est; et haec autem nostra eadem est, quae ex illa est.* (In other authors cf. Plin. *Nat*. 2.80 *et ipsa autem luna*, 5.26 *et terra autem*, etc.; Quint. *Inst*. 4.1.66 *et Demosthenes autem ... et M. Tullius ...*) In our passage *et* means 'also' and the *autem* adds an afterthought by way of explanation, cf. *OLD* s.v. 4a.

6.20.6 potest terram mouere impressio spiritus: fortasse enim aër et extrinsecus alio intrante aëre agitatur, fortasse aliqua parte subito decidente percutitur, et inde motum capit.

enim aer et Zθπ: enim aer δ: in ima (*uel* imo) aër *Gercke*

The *et* is part of the paradosis, but is puzzling; perhaps δ is correct, for it could be a dittography of *-er* in *aër*. But Gercke saw that we could do with being told that this air is beneath the earth. Maybe there is a lacuna, e.g. *fortasse enim aër <sub terris congregatur>, et extrinsecus ...*

6.21.1 Theren et Therasiam et hanc nostrae aetatis insulam spectantibus nobis in Aegaeo mari natam ...

nobis Ω: nautis *Albertus Magnus Meteor. 3.2.19, ut coni. Pincianus*

It is doubtful whether *spectantibus nobis* can mean 'et sous nos yeux' (Oltramare) or the like without implying that S. himself was there (cf. Caes. *Gal.* 7.25.1 *accidit inspectantibus nobis quod dignum memoria uisum praetereundum non existimauimus*; Cic. *Phil.* 1.24 *eas leges quas ipse nobis inspectantibus recitauit*); and even if it could, is it really likely that many of S.'s contemporaries saw the event in a rather obscure corner of the Aegean? *nautis* must be right: for sailors at the eruption of Thia see Justin *Hist. Phil.* 30.4.1-2, and for sailors at another submarine eruption, Str. 6.2.11, C277.

6.22.1 si quando magna onera per uices uehiculorum plurium tracta sunt, et rotae maiore nisu in salebras inciderunt, terram concuti senties.

uices Ω: uicos ς: uias *Gertz*[2]: axes *Schultess*[2] uehiculorum plurium *ante* per *transp. Rossbach*[3] *942* plurium BVθPW: plurimum ZR: plurima U: *om.* A: plurium <ordine> *Fromondus* sunt Ω: eunt *Gercke* terram Ψ: tecta ZL[2], *ut coni. Gertz*[2]

Usually *per uices uehiculorum plurium* is taken to mean 'par plusieurs véhicules avançant en file les uns derrière les autres' (Oltramare) vel sim. But this sense of *uices* is hard to parallel. *per uices* on its own regularly means 'alternately, turn and turn about' (*OLD* s.v. *uicis* 1b), so if our phrase meant anything it ought to be something like 'by relays of several wagons', which is not what we want. At S. *Ep.* 108.20 *nec tantum caelestia per certos circuitus uerti (credis), sed animalia quoque per uices ire et animos per orbem agi,* we find *per uices* meaning 'through successive states'; but (*pace OLD* s.v. *uicis* 7) this does not show that our passage could mean 'a series of linked wagons'; it ought to mean a succession of different wagons. So *per uicos* is attractive, and then one must either change to *uehiculis pluribus* (on the assumption that this was changed to the genitive after the corruption *uices* appeared) or an ablative must be inserted, e.g. Fromondus's *<ordine>* or *<serie>* (for the sake of rhythm I prefer the word order *<ordine> uehiculorum or uehiculorum <ordine> plurium tracta sunt*); cf. *Ep.* 90.9 *nec in hunc usum pinus aut abies deferebatur longo uehiculorum ordine uicis intrementibus.*

ZL[2] confirm Gertz's conjecture *tecta.*

6.22.2 et sic commouetur omne tectum cauatae uallis.

et Ψ: ac Z cauatae Ψ: cauae Z

For *caua uallis* and similar phrases see *TLL* 3.716.14-20 (including *Oed.* 569). The adjective *cauatus* is not found elsewhere in S., which is no strong objection, but rhythm favours *cauae.*

6.23.3-4 omnia licet antiqua ducum regumque exempla transierit, ex his quae fecit nihil tam magnum erit quam scelus. hic Callisthenes in libris quibus describit quemadmodum Helice Burisque mersae sint, quis illas casus in mare ...

scelus. hic Ω: scelus hoc. *Fickert*: scelus <hoc>. hic *Michaelis*[2]: <tale> scelus. hic *Gercke in app.* libris Ψ: his libris Z

If we need a demonstrative with *scelus*, I would prefer *quam <hoc> scelus. hic* ..., for with Fickert's and Michaelis's conjectures, the position of *hoc* is highly unusual, and the rhythm less good. But, as Professor Watt points out, the paradosis can mean '... nothing will be as great as a crime', which is possible sense, and indeed the addition of a demonstrative might misleadingly suggest a contrast with other crimes of Alexander. Before *libris* one might alter Z's *his* to *iis*, but the rhythm is better without either.

6.24.4 'immo' inquit 'ceu cum frigore inhorruimus, ...

ceu (seu W) Ψ: quemadmodum Z

ceu occurs elsewhere in S. only at *Tro.* 20 (and at *Oct.* 191). Previously the word was poetic, and otherwise first appears in prose in the Elder Pliny (where it is frequent), and later in Suetonius and Apuleius. The arguments are all on its side here: *ceu* could easily have been omitted before *cum*, or could have been glossed by *quemadmodum*, but it could not easily have displaced an original *quemadmodum*.

6.25.1 haec (sc. latera terrae) nonnumquam adeo concutiuntur ut aedificia superposita procumbant, nonnumquam in tantum ut parietes ... decidant ...

adeo (a deo R) θπ: in tantum Z: *plura om.* δ

Z's *in tantum* was vulnerable to omission after *nonnumquam*, and *adeo* was an easy supplement. It is less likely that the subsequent *in tantum* was consciously or unconsciously repeated, ousting an original *adeo*.

6.26.4 Cypron ambit altum mare, et agitatur; Tyros et ipsa tam mouetur quam diluitur.

agitat *Gercke*   alluitur *Skutsch*

Gercke's *agitat* is unnecessary, and spoils the rhythm, *Cypron āmbĭt āltūm mărĕ/, ĕt ăgĭtātŭr,* two cola, with hiatus after *mare.*

diluitur would suggest Tyre was washed away (cf. Plin. *Nat.* 33.76 *ita profluens terra in mare labitur ruptusque mons diluitur*). Skutsch's *adluitur* is excellent, cf. *Dial.* 12.7.2 *totum Italiae latus quod infero mari adluitur maior Graecia fuit.* Tyre was originally an island that was joined to the mainland by a mole (Str. 16.2.23, C756-7, Mela 1.66).

6.27.3 crassae itaque (aquae) et graui caligine sempiternaque tectae nihil nisi pestiferum in se et corporibus nostris contrarium habent. aër quoque ...

habent Ψ: *om.* Z

The indifferent clausula creates a suspicion that *habent*, omitted by Z, is either in the wrong place in Ψ (after *in se* is a possible alternative) or is a stop-gap (*continent* would give better rhythm and was easily omitted after *contrarium*).

6.29.1 nam quod aliquos insanis attonitisque similes discurrere, fecit metus, qui excutit mentes, ubi priuatus ac modicus est; quid, ubi publice terret, ubi cadunt urbes, populi opprimuntur, terra concutitur? quid mirum est animos inter dolorem et metum destitutos aberrasse?

quod *om.* OT   aliquos Z^{c}θπ: *quid* Z^{1}, *incertum* (*fort.* aliquod): aliquot *Erasmus*^{2}

With the paradosis *aliquos*, as Oltramare edn., 286, n. 2, says, the sentence must be articulated thus: *nam quod aliquos insanis attonitisque similes discurrere fecit metus (qui excutit mentes, ubi priuatus ac modicus est; quid, ubi publice terret, ubi cadunt urbes, populi opprimuntur, terra concutitur?), quid mirum est animos inter dolorem et metum destitutos aberrasse?* This is rather clumsy (though it is accepted by Setaioli 15). Erasmus' *aliquot*, taking *discurrere* as a perfect, gives a neater structure.

Vottero, however (edn., 191-2) objects that (a) *aliquot* is rare in S. (as a pronoun only at *Ben.* 2.8.1 *nam aliquot postea, qui idem rogarent, inuenti sunt* [*aliquod* N[1]R[1]: *aliqui* M]); (b) with *facere quod* + indicative in prose there is never a change of subject between the two clauses, except where one verb is impersonal (he refers to *TLL* 6.106.84-107.12). Vottero omits *quod* and takes *discurrere* to be an infinitive. But point (a) is weak (indeed Setaioli points out that this would be the only case of *facere* plus accus. and infin. in Seneca, so Vottero's text is equally open to such objection). Point (b) is not much better: change of subject is found at Mart. 10.13(20).1-5 *Ducit ad auriferas quod me Salo Celtiber oras, pendula quod patriae uisere tecta libet, ... tu facis*, where notice that, as in our passage, the *quod*-clause precedes. Furthermore Vottero ignores the context: the preceding chapter has spoken of harmful *spiritus* coming from below the earth. At the start of this chapter there is a contrast: the people who behaved as though they were mad in this recent earthquake were affected not by *spiritus* but by fear. The contrast is plain with *nam quod ... discurrere, fecit metus ...*, but Vottero's text loses the contrast.

It is true that *aliquot* would be unusual for S., but it should not be ruled out; if change were needed, one could alter *aliquos* to *aliqui*, supposing that *aliquos* arose when *discurrere* was assumed to be an infinitive.

6.30.2 uides totas regiones a suis sedibus reuelli, et trans mare iacere quod in confinio fuerat, uides et urbium fieri gentiumque discidium, cum pars naturae concita est, et aliquo mare, ignem, spiritum impegit, quorum mira ut ex toto uis est; quamuis enim parte saeuiat, mundi tamen uiribus saeuit.

cum pars naturae concita est *om.* θ   et (*om.* W) aliquo Zδθυ: deesse et aliquo P: dicere et alio R: *ante* et *spatium relinquit* U

deesse has prompted numerous conjectures, but its authority is doubtful: it goes back to ρ (on the assumption that R's *dicere* is just a corruption), perhaps even to π, if U's gap shows there was something in its exemplar; and in view of θ's omission of the preceding words, we cannot be certain whether π or δ represents Ψ more accurately; δ or Z could have omitted a troublesome word deliberately.

If *deesse* is the paradosis, none of the conjectures convinces: *de se* (Fortunatus), *ex se* (Muretus), *e sede* (Madvig[2]), *excessit* (Leo), *desaeuit* (Rossbach[5]), *dehiscens* (Vottero[3]); and Gercke proposed *concita* (or *concitae*) *<ausa> est deesse.* So it should be omitted, given its flimsy authority. (Could *deest* have been used like *uacat* to indicate a lacuna at an early stage of the tradition? Cf. Havet p. 378 §1513 for speculations about manuscript variants caused by contracted *deest.*)

6.32.3 ingenti itaque animo mors prouocanda est, siue nos aequo uastoque impetu adgreditur, siue cotidiano et uulgari exitu.

aequo] caeco *Haupt[1] 337*: saeuo *Koeler*

In Koeler's posthumous edition *saluo* is clearly a misprint for *saeuo* (he cites Hor. *Carm.* 3.1.27-8 *nec saeuus Arcturi cadentis impetus* as a parallel). *aequo* cannot be defended as meaning 'impartial', with reference to the commonplace of death the equaliser, because a contrast is needed with *cotidiano et uulgari exitu*, which is also *aequus* (nor can *aequo* mean, as Oltramare and Alexander think, 'affecting everybody, i.e. causing large-scale suffering'). *saeuo* is easier palaeographically than *caeco.*

6.32.3 nihil refert quam minax ueniat, quantumque sit quod in nos trahat; quod a nobis petit, minimum est; hoc senectus a nobis ablatura est, hoc auriculae dolor, hoc umoris in nobis corrupti abundantia, hoc cibus parum obsequens stomacho, hoc pes leuiter offensus.

a nobis (*post* senectus) *om.* Z   in nobis Ψ: intra nos Z

S. generally (over 20 times in prose and verse) uses the dative, not *a(b)*, with *aufero*; *a(b)* is common in other authors (cf. *TLL* 2.1329.15-21, including Plautus and Cicero), but in S. is found only at *Ben.* 5.21.3 ... *nec ab ullo beneficium repetam, nisi a quo non ablaturus ero, sed recepturus*, but there *recepturus* requires *a quo.* Here Ψ's *a nobis* is unnecessary, and contrary to S.'s normal practice, so Z should be preferred. After *umoris*, there is little to choose between Ψ and Z.

6.32.5 siue illam insidiae siue morbi petent, siue hostium siue ciuium gladii, siue insularum cadentium fragor, siue ipsarum ruina terrarum, siue uasta uis ignium urbes agrosque pari clade complexa, qui uolet †illam eam† accipiat.

morbi - gladii siue *om.* Z[1], *add. in mg.* Z[c] siue ciuium Z[c]: *om.* Ψ qui uolet illam] cum libet illi *Erasmus*[1] illam eam ZθPW: illa eam U: eam δR: illam Cam.DE: illam, a me *Gruterus*: illam casus *Gertz*[2]

The paradosis is *illam eam*. Recent editors accept δ's *illam*, and understand 'laissons-la prendre à qui la voudra' (Oltramare, and similarly Codoñer Merino and Vottero) or 'let life be taken by whoever or whatever wants it' (Corcoran—but 'or whatever' is not in the Latin). This is difficult, for most of the *siue*-clauses describe natural forces, not people, that threaten one's life, and *accipiat* is a mild word for the action of these powerful forces. But maybe Gertz is right: he offered *casus* as a supplement, but *eam* could be a corruption of it (*casus* → *cas* → *eas* → *eam*). The point of *accipiat* could be that the *anima* puts up no resistance but lets itself be taken. For *qui uolet ... casus*, cf. 6.17.1 *quod semper moueri uult*, and 2.24.2 *itaque non est quod eum spectes cuiusque rei habitum qui illi non ex uoluntate est.*

6.32.6 nec rogaueris nec timueris nec te uelut in aliquod malum exiturum tuleris retro.

exiturum θπ: exituram ZV: exiturus AB: exitura *Bongarsius*

Bongarsius' conjecture should be accepted, on the assumption that the *anima* is still being addressed; for the rhythm is better, and Z's *exituram* supports it (V's reading is probably an accident).

6.32.10 quid, oro te, refert num perueniam ad illos?

num per- θ: non per- Z: nuper ρ: si non per- δ: unde per- υ

There is no other example of *refert num* in S., nor in any other author that I know of (nor of *interest num*, cf. *TLL* 7.1.2285.63-2287.10). Is *num* a corruption of the normal *an* (cf. e.g. *Dial.* 5.27.2, 8.6.1, *Ben.* 5.20.2, *Nat.* 2.34.4)?

8

Notes on *Naturales Quaestiones* Book 7

7.1.3 ... quod (sol) lunam numquam nec implet nisi aduersam sibi nec obscurat <***>

nec (*ante* implet) Z: *om.* Ψ   nec obscurat] nisi uicinam obscurat *Alexander*   <nisi uicinam> *post* obscurat *suppl. Gercke*: <nisi obliquam> *Oltramare*: <nisi in coniunctione> *Codoñer Merino*: <nisi transuersam> *André 197*

Z's first *nec* should probably be accepted. Gercke's supplement gives poor rhythm, Oltramare's and André's less good sense. Codoñer Merino's is possible; so, for example, is <*nisi propinquam*> or <*nisi adiacentem*> (cf. S. *Nat.* 1.10, Iul. Val. 1.4).

7.2.2 illa quoque possunt et tenui constare materia, nec ob hoc discuti adsiduo caeli circumactu.

et tenui θπ: extenui Z: tenui δ

S. regularly uses *constare* = 'be composed of' both with *ex* (*TLL* 4.527.71-3) and with simple ablative (ibid. 528.24-6). Here *et* ... *nec* ... is highly appropriate.

7.5.1-2 Vt a proximis mendaciis incipiam, falsum est trabes et faces exprimi turbine: turbo enim circa terras concipitur ...; at contra trabes editior caeli pars ostentat, itaque numquam nubibus obstiterunt. [2] praeterea turbo omni nube uelocius rapitur, et in orbem uertitur, super ista celeriter desinit, et ipse se ui sua rumpit. trabes autem non transcurrunt nec praeteruolant ut faces, sed commorantur et in eadem caeli parte conlucent.

trabes et faces ZF: faces (fasces A) et trabes δHπ   celeriter Z: uelociter Ψ *celeriter* and *uelociter* give equally good sense and rhythm, but *uelocius* just before could have caused corruption of *celeriter.*

Gemoll[2] proposed deleting *ut* before *faces* in §2 (*ut* being a repetition of the end of *praeteruolant*), because at the start of §1 S. says it is not true that *trabes et faces* are produced by a *turbo*, yet then proceeds to talk solely about *trabes*, and indeed here in §2 seems to differentiate between the two, saying that *faces praeteruolant* and *trabes* do not. But this contrast is consistent with S.'s use of the terms elsewhere, where *faces* are a form of shooting-star, fast moving and short-lived (cf. e.g. 1.1.11 *sic faces quoque transcurrunt et interdiu*, 6.3.3 *actae in transuersum faces* are listed separately from *crinita sidera*, 7.23.2), whereas *trabes* are hard to distinguish from comets (1.15.4, 7.5.5), and can last many days (7.5.3). So Gemoll's deletion of *ut* is wrong, but his diagnosis of a problem is sound. It could be loose writing by S., but maybe in §1 we should supply *at contra trabes <et faces> editior ...*, for this first argument applies to *faces* as well as *trabes* (both occur high above the atmosphere); it is then followed in §2 by an argument about *trabes* only.

7.6.2 potest enim spiritus per angusta elisus accendere supra se positum aëra plenum alimentis idoneis igni, deinde propellere †ex nitido nec† ex aliqua causa refluat rursus ac remittatur ...

ex nitido nec Z: ex nitido ne Ψ: ex humili donec *Koeler*: ea niti, donec *Madvig*[2]: eum niti, donec *Gercke*: exin niti donec *Garrod*[2]: et eniti donec *Rehm*[2] *10 n. 1*: et niti donec *Oltramare*: [ex] niti donec *Castiglioni ap. Oltramare*

niti is not found with the infinitive in S. (nor *eniti*), so Madvig's, Gercke's and Castiglioni's conjectures are suspect. The object of *propellere* can readily be understood, so, *pace* Madvig, it is not necessary to supply one. Against Garrod, S. does not use *exin* or *exinde*. With Oltramare's text, *propellere et niti*, one must not (like Corcoran and Codoñer Merino) think *niti* is transitive; Oltramare more wisely translates 'puis la pousser devant lui et faire effort' (similarly Vottero). But then *niti* is padding, for *propellere* already contains the idea of considerable force. Koeler's conjecture is best, supplying a useful indication of direction, and being palaeographically plausible (*ex (h)ūili donec* → *ex nitido nec*); for *ex humili* cf. 1.8.3, 7.8.4, *Dial.* 11.9.8, *Ep.* 77.2.

7.7.1 Aduersus haec multa dicuntur.

hoc ZL^2

At 7.13.1, editors since Fickert print *Aduersus haec ab Artemidoro illa dicuntur*, but the paradosis is *aduersus hoc* [*hec* BR]; and at 7.18.1 *Aduersus hoc protinus respondetur ...* is better attested (*hoc* ZΘP: *hec* αU: quid W^1, incertum). Elsewhere too S. uses *aduersus hoc* with *respondeo* and the like (*Dial.* 8.6.1, *Ben.* 6.13.4, 6.15.2, *Nat.* 3.7.1, *Ep.* 85.31, 87.12, 87.26). So Z's reading is preferable here. (Livy is the only writer known to me who uses *aduersus haec* in such contexts: 4.41.1, 8.2.12, 30.31.1, 36.32.3).

7.8.3 num quis nostrum crederet lunam aut quinque sidera rapi uento aut turbine rotari?

Pincianus missed mention of the sun. His *lumina* (supposedly meaning sun and moon) for *lunam* is not persuasive, but maybe supply *lunam aut <solem aut>*. In Latin *sol* commonly comes before *luna* when they are paired, but for the reverse order compare Cic. *Tusc.* 1.63 *lunae solis quinque errantium motus.*

7.9.3 praeterea uiolentius celeriusque in terra circaque eam uoluitur.

Vottero follows Castiglioni^2, who changed *eam* to *terram* on the grounds that *circaque eam* is feeble and not in S.'s manner. This is a clear case of an unnecessary conjecture spoiling a clausula.

7.11.1 quas (opiniones) antequam exponere incipiam, illud in primis praesumendum est.

incipio Z

The subjunctive after *antequam* is much more frequent than the indicative in Seneca, and in post-Augustan and late Latin generally (H-Sz 600), so corruption was more likely to proceed from indicative to subjunctive. Z's reading is supported by *Ep.* 117.4 *aduersus hos quid ab aliis respondeatur audi, antequam ego incipio sedere et in alia parte considere.*

7.11.2 †forma eius non est una†: quamuis enim Graeci discrimina fecerint eorum, ..., tamen omnes isti eiusdem notae sunt cometaeque recte dicuntur.

<unum est genus,> forma eius non est una *Gercke in app.* eius non Zθπ, A *u.l.*: cometarum non A^{1}BV: eius [non] *Fortunatus*: eis [non] *Erasmus*[2]: eis non *Muretus*: eis uero *Haase*: eis, ut nomen *Leo*: oculis nota *Schultess*[2]: non setius *Gercke in app.*: sola non *Brakman*[1] eis <si> non est una, <nota est una> *Michaelis*[2]: eis non est <nisi> una *Rossbach*[3] *942*: eiusdem notae est, una *Castiglioni*[2]

Leo's conjecture is accepted by Oltramare and more recent editors. But does S. really want to say that the *forma* of these different comets is one? Surely we expect 'even though their forms are different, they are really all the same kind of thing'. Most of the other conjectures also give the wrong sense. Gercke's suggestion of a lacuna is maybe on the right lines, though the *genus/forma* contrast does not seem right. Maybe one should restore something like *forma eis non est una, <sed una natura est>* or *<sed eiusdem notae* [used just below] *sunt>*; or *<est una natura>* (Professor Watt). For the difference between *natura* and *forma* (of a person), cf. Cic. *Tusc.* 4.80 *qui se naturam cuiusque ex forma perspicere profitebatur.*

7.11.3 ... prout cuique acrior acies aut hebetior est, ita ait aut lucidiorem esse aut rubicundiorem et crines aut in interiora deductos aut in latera demissos.

deductos ZθPυ: did- R: demissos δ: reductos *Gronouius*[1] demissos Zθπ: deductos δ: dimissos O, *ut coni. Fickert*

According to the sharpness of their eyesight, different people will see different amounts of detail in a comet. The penultimate phrase must mean (in Professor Nisbet's words) 'if one's sight is *hebetior*, the "hairs" of the comet seem to coalesce with the core of the tail (*interiora*)', whereas sharper sight will detect hairs spreading outwards as well. In this case *deductos* is probably acceptable, in the sense 'draw out, stretch out' (*OLD* s. *deduco* 3a, b). Gronovius's *reductos* makes the sense clearer, but spoils the rhythm, although the heroic clausula is less objectionable in mid-sentence. Another possibility is *restrictos*, though that is further from the paradosis (cf. 7.17.1 *non in rotundum restricta*). In the last phrase *dimissos* is better sense than *demissos*, and is necessary to create a clear enough contrast if we keep *deductos*.

7.12.5 quomodo ergo potest altera stella usque ad alteram stellam ignem mittere, ita ut utraque iuncta uideatur, cum sint ingenti regione diductae?

stellam δθυ: *om.* Zρ

It would be strange if any interpolator thought *stellam* needed to be supplied, so perhaps it is genuine, for it was easily omitted and it improves the clausula. Professor Nisbet suggests one might omit the first *stella*. (Compare perhaps Front. *Ant.* 2.16.1 V.d.H.[2] (2.13.1 V.d.H.[1]) *tum pastor unus ad alterum pastorem ...*).

7.13.2 nam si illi credimus, summa caeli ora solidissima est, in modum tecti durata et alti crassique corporis, quod atomi congestae coaceruataeque fecerunt.

Shackleton Bailey says that he cannot see how *corpus* can be *altum*, and proposes *alta*. He presumably means to take *in modum tecti durata et alta* together, but how could S. have any quarrel with Artemidorus for saying that the outer heaven is *alta*? We need some other objectionable feature of Artemidorus' theory, and *alti* is intelligible, describing the thickness or depth of the heavens' outer shell: *crassi* means dense as opposed to rarefied, *alti* means thick as opposed to thin. Cf. Scrib. Larg. 240 *crustam altam*, Stat. *Theb*. 11.502 *alto ... auro* (of a shield), and other passages at *TLL* 1.1775.47-68.

7.13.3 habet tamen spiramenta quaedam et quasi fenestras, per quas ex parte exteriore mundi influant ignes, non tam magni ut interiora conturbent <***> rursus ex mundo in exteriora labuntur.

habent *Usener ap. G. Müller 44-5* rursus Zθπ: qui rursus δ: <sed> qui rursus *Gercke*: rursus<que> *Castiglioni*[2] labantur *Gercke in app.*

δ's *qui* has no authority. I find this sentence obscure, and think there could be a larger lacuna, for I would expect Artemidorus to have used the *fenestrae* to explain the fixed stars as well as comets.

7.14.1 Soluere ista quid aliud est quam manum exercere et in uentum iactare bracchia?

manum] inane *Gruterus* exerere B[c], *ut coni. Skutsch*

Recent editors retain *exercere*, but *manum exercere* is a good thing to do, so in that case we need something like 'in vain' with the phrase. One could supply *frustra* or the like; or one could transpose *in uentum* to follow *quam*, but though this may be palaeographically plausible, *manum exercere et iactare bracchia* is a weak pair of phrases. *exerere* seems best: 'sticking out your fists and punching the wind'. (S. uses *manum exercere* at *Dial.* 5.18.2, 8.6.3, *exerere* at *Ben.* 2.31.1.)

7.14.2 fieri enim non potest ut tanta uis ponderis, quantam ille substituit, pendeat et leuibus innixa sit.

Garrod[2] conjectured <*animo*> *ille* on the grounds that there was no example of *substituo* meaning 'suppose' without the addition of some such word. Vottero accepts Garrod's objection but prefers B's *sustinuit.* But *OLD* s. *substituo* 2 provides parallels: V. Max. 9.11.intro. *uitae humanae cum bona tum etiam mala substitutis exemplorum imaginibus persequimur*, Quint. *Inst.* 3.8.54 *utimur ficta personarum, quas ipsi substituimus, oratione.*

7.14.4 aut si hoc sine ullo probabili argumento licet respondere, quid est quare non aliquis aut omnes stellas moueri dicat aut nullam?

Gertz[2] correctly saw that *respondere* cannot mean 'affirm', which is the required sense. His *spondere* should be accepted. Alexander proposed *responde*, which is accepted by Vottero. But Alexander does not explain what *si hoc sine ullo probabili argumento licet* then means, and Vottero's translation is impossible: 'se ciò non è possibile per mancanza di qualsiasi argumento accettabile'.

7.15.1 primo igneus ac rubicundus orbis fuit, clarumque lumen emittens, quanto uinceret noctem, deinde paulatim magnitudo eius districta est, et euanuit claritas, nouissime totus intercidit.

adstricta ZL^2

With *districta* the meaning must be that the comet's size got bigger and fainter (so Vottero), not smaller and fainter (as Oltramare, Corcoran, Codoñer Merino) (cf. *OLD* s.v. *distringo* 1 'to ... stretch out or apart'). But comets are usually brighter when they are larger, i.e. closer to earth (cf. c. 17.3 fin.). *adstricta* is possible, meaning 'contracted, compressed' (cf. *TLL* 2.960.64-961.7; S. *Ep.* 106.5 *an frontem adstringant*; Verg. *G.* 1.91 *uenas astringit hiantis*; Quint. *Inst.* 11.3.81 *labra ... astringuntur et diducuntur*; *Aetna* 561 *spiritus adstrictis elisus faucibus*). *adstricta* improves the rhythm too.

7.19.1 per repercussionem uicinorum siderum ...

repercussus Z

Z's text should be accepted and *repercussio* removed from dictionaries of classical Latin, where it occurs nowhere else. It is found in medieval Latin (Du Cange s.v.) and glossaries (*CGL* 2.172.41, 230.51).

7.21.4 uidelicet ille umidiora habuit et aptiora ignibus, quae persecutus est, huic rursus uberior fuit et plenior regio, huc itaque descendit inuitante materia, non itinere.

ille Zδθ: illo Rυ: *om.* P: illa *uel* ille ea *Gercke in app.* (h)umidiora ZαHP^{c}υ: humiliora FP^1: fumidiora *Oltramare* humiliora ... habuit <inania et altiora> aptiora ... et plenior <inferior> regio *Shackleton Bailey* ignibus que αθPU: ignibusque Z: ignibus quidem W^1, *ut uid.* uberior Ψ: haec uberior Z: uberior haec L^2

With Z(L^2)'s *haec uberior*, we simply need to supply e.g. *ille <illic>* and adopt Oltramare's *fumidiora* to have satisfactory sense: *illic* and *haec* refer back to the descriptions of the paths of the two comets in the previous sentence. For *ille <illic>* cf. *Ben.* 3.29.4 *adspice Rhenum, adspice Euphraten, omnes denique inclutos amnes: quid sunt, si illos illic, unde effluunt, aestimes?*, *Ep.* 95.2 ... *ab iis qui illum ommutescere illic* [*ilico* Erasmus] *cupiunt*. For *fumidus* in this context, of materials fuelling fire, cf. 2.54.1 *e terra terrenisque omnibus pars umida efflatur, pars sicca et fumida; haec fulminibus alimentum est, illa imbribus*, 2.57.3, Plin. *Nat.*

2.111 ... *umidam a terra, alias uero propter uaporem fumidam exhalari caliginem certum est.*

7.21.4 omnibus autem stellis in eandem partem cursus est, id est contrarius mundo.

Nothing in the context makes clear that *stellis* means planets, as opposed to stars in general, and indeed *omnibus* suggests the latter, of which the statement is not true. Gercke supplied <*V*> *stellis*, but *omnibus quinque stellis* sounds odd, because there are many more than five *stellae* all told; so maybe *stellis* <*erraticis*> is preferable.

7.22.1 fieri non potest ut ignis certus in corpore uago sedeat, et tam pertinaciter haereat quam quae natura ne umquam excuterentur aptauit.

qu(a)e ZBVθPU: quia A: quod W^1: *om.* R: quem *Fortunati cod.* excuterentur Zδθπ: -etur R, *Fortunati cod.*

Gercke and more recent editors accept Fortunatus' variant, where the antecedent of *quem* must be *ignis*; but no change is needed. Oltramare mistranslates *quem ... excuteretur*: 'Il est impossible qu'un feu reste fixe dans un milieu instable et s'y maintienne aussi invariablement que doit le faire *un* corps disposé par la nature de manière à ne jamais perdre son assiette' (my emphasis); Corcoran and Codoñer Merino translate similarly, but Vottero is accurate ('... così tenacemente come quello che la natura ha allestito ...'). The sense that Oltramare and others want is appropriate, and is given by the neuter *quae* of the paradosis: the reference of the *quae*-clause is to stars and planets.

7.23.1 nullis ignibus ordinariis et caelestibus iter flexum est: sideris proprium est ducere orbem.

ducere Ψ: scribere Z

ducere is found in this sense (cf. e.g. Ou. *Met.* 8.249 (of compasses) *altera pars staret, pars altera duceret orbem*; Manil. 1.675 *sic per tris gyros inflexus ducitur orbis*; Sen. *Contr.* 1.5.2; S. *Pha.* 973-4 *uasti pondera mundi / librata suos ducunt orbes*), but could be influenced by *ducit* three or four lines above, in a different sense. However, I can give no close parallel for *scribere* in this context, unless it is an extension from the geometrical use (Cic. *Tusc.* 5.113 ... *unde, quo quamque lineam scriberent*, S. *Ep.* 74.27 *utrum maiorem an minorem circulum scribas*). Since *ducere* could have intruded from above, *scribere* should be accepted.

7.24.1 (sidera) alios et alios circulos habent.

alia [set] alios *Gercke in app.*

alius atque alius is the normal expression in classical Latin, and in S.; but there is no need to emend, for S. also has one case of *alius aliusque* (*Ep.* 88.34 *in alias animalium formas aliasque coniectus*—note the rhythm; also found in other early imperial writers, see *TLL* 1.1640.42-53), and *alius et alius* is later found in Apuleius, jurists and others (*TLL* 1.1640.54-8).

7.27.5 ... quaedam eduxit a turba, ut singula et conspicua procederent, quaedam in gregem misit.

educere a(b) is unusual—never found in S., and rare elsewhere (and then in contexts where *ex* would give a different sense: Caes. *Ciu.* 3.30.3 *ex castris statiuis a flumine Apso exercitum educunt*; *ab urbe* is found at Liu. 3.21.2, Suet. *Aug.* 89.1; *a palatio* SHA *Comm.* 4.5). Also, in *TLL* 5.2.123.35-7 this is the only instance of the verb meaning 'Translate: i.q. separare'; but S. seems to stretch the meaning of the verb elsewhere, for *TLL* ib. 37-9 also gives *Ep.* 66.40 'i.q. promouere' and 104.13 'i.q. expellere', both passages also on their own. Nevertheless, we should maybe read *<s>eduxit a* (cf. c. 25.7 *cur tam seducti a ceteris errent*) or *eduxit e*.

7.28.1 non iudicas sidus esse quod futura denuntiat?

denuntiat (-iet R) Ψ, Z *uaria lectio s. l.*: demonstrat Z[1]

denuntio is common in this sense, of one thing being a sign or omen of another (*TLL* 5.1.555.20-62; cf. *Nat.* 2.34.1, 2.47, 6.26.3, and *denuntiatio* at 1.1.3), *demonstro* is rare: *OLD* s. v. 3a and *TLL* 5.1.508.12-5 give only Plin. *Nat.* 18.354 *cum ab austro uel coro aut fauonio nocte serena fulgurabit, uentum et imbrem ex isdem regionibus demonstrabit*, 25.19. So here the rarer *demonstrat* is right: the gloss *denuntiat* appears as a variant in Z, as the text in Ψ.

7.29.1 ... iter suum lentissime efficit ...

efficit Ψ: conficit Z

iter conficere is common, cf. *TLL* 4.197.25-43, including S. *Dial.* 7.23.4; *iter efficere* is not common: *TLL* 5.2.165.69-72 gives Caes. *Ciu.* 3.102.1 *quantumcumque itineris equitatu efficere poterat cotidie progrediebatur*, Ou. *Met.* 6.519 *iamque iter effectum.* Although it is not found elsewhere in S., the rarer phrase should probably be preferred.

7.29.3 Haec sunt quae aut alios mouere ad cometas pertinentia aut me.

[ad] cometas pertinentia (pertintia R) *hic habent* RθP[1]U, cometis pertinentia *hic* L[2]P[c]: ad cometas pertinentia *post* sunt Z, *post* quae δ

This is probably the best place for *ad cometas pertinentia*, since it helps the clausula, but in view of the different word order in Z, could it be a gloss? *Haec* could simply mean 'what precedes', as at 2.11.3 *Haec necessarium fuit praeloqui ...*, 6.32.1 *Haec, Lucili uirorum optime, quantum ad ipsas causas: illa nunc quae ad confirmationem animorum pertinent.*

7.30.2 (tractandum est) an aeque omnis caeli regio idonea in qua creentur (cometae), an quacumque ire ibi etiam concipi possint, et cetera.

concipi (*plura om.* B: -cepi W) Ω: conspici *Bentley*

With *concipi* ('be formed') this clause virtually repeats the last; Bentley's conjecture is excellent, as is another of his at ...

7.30.4 ... siue in sanctiore secessu maiestas tanta delituit, et regnum suum, id est se, regit, nec ulli dat aditum nisi animo.

regit (freg- U) Ω: tegit *Bentley*

Bentley is clearly right, for S. in the context is stressing how much of nature is still unknown to us.

7.32.1 ... cum aliquis pluuius interuenit dies, quem perdere libet.

Alexander proposed the transposition *libet perdere*, which greatly improves the clausula. However, I suspect that S. avoided collocations like *quem libet*, which could be mistaken for *quemlibet* (though note Hor. *Ep*. 1.6.53 *cui libet hic fasces dabit*).

7.32.3 in hoc uiri, in hoc feminae tripudiant; mares inter se uxoresque contendunt uter det latus mollius.

mares] maritus *Leo*: mariti *Gercke in app.* uxor[es]que *Leo*

If the sense must be 'husbands and wives', we need to emend; but the paradosis can stand. This kind of dancing may be acceptable among women of low social status, but is shameful for all males, whatever their marital status, and among women it is particularly shameful for the married. (*TLL* 8.422.66 says *mares = mariti* at Vict. *Caes*. 4.7, but I would not attribute that sense to *mares* here.)

Index of Passages Discussed

Passages of the *Naturales Quaestiones*

Passages of other works

Index Rerum et Verborum